T0073756

Evolution and the Machinery of Chance

Evolution and the
Machinery of Chance

bability, and

:e in Biology

ABRAMS

The University of Chicago Press
Chicago and London

The University of Chicago Press, Chicago 60637
The University of Chicago Press, Ltd., London
© 2023 by The University of Chicago
All rights reserved. No part of this book may be used or reproduced in any manner
whatsoever without written permission, except in the case of brief quotations in critical
articles and reviews. For more information, contact the University of Chicago Press,
1427 E. 60th St., Chicago, IL 60637.
Published 2023
Printed in the United States of America

32 31 30 29 28 27 26 25 24 23   1 2 3 4 5

ISBN-13: 978-0-226-82661-5 (cloth)
ISBN-13: 978-0-226-82663-9 (paper)
ISBN-13: 978-0-226-82662-2 (e-book)
DOI: https://doi.org/10.7208/chicago/9780226826622.001.0001

Library of Congress Cataloging-in-Publication Data

Names: Abrams, Marshall (Professor of philosophy), author.
Title: Evolution and the machinery of chance : philosophy, probability,
    and scientific practice in biology / Marshall Abrams.
Description: Chicago : The University of Chicago Press, 2023. |
    Includes bibliographical references and index.
Identifiers: LCCN 2022045069 | ISBN 9780226826615 (cloth) |
    ISBN 9780226826639 (paperback) | ISBN 9780226826622 (ebook)
Subjects: LCSH: Evolution (Biology)—Philosophy. | Natural selection. |
    Biological fitness. | Population—Environmental aspects.
Classification: LCC QH360.5 .A27 2023 | DDC 576.8/2—dc23/eng/20221115
LC record available at https://lccn.loc.gov/2022045069

♾ This paper meets the requirements of ANSI/NISO Z39.48-1992 (Permanence of Paper).

In memory of Lois Barnett
May 20, 1933–June 17, 2016

Contents

Preface

The diversity and quantity of the evidence for the theory of evolution is unparalleled, and its value in guiding research in biological science has been shown repeatedly. Despite this, there are puzzles about the foundations of evolution theory. Though occasionally troubling to scientists, these puzzles almost never impede scientific progress. Microevolutionary processes—those that take place within biological populations over months or millennia rather than millions of years—raise particularly interesting issues. Philosophers of biology and some biologists have, I believe, missed important insights about the nature of these evolutionary processes by focusing too closely on what is directly observable in evolving populations. Insufficient attention has been paid to the fact that much of the best research in evolutionary biology uses modeling and statistical inference to make judgments about hidden causal patterns.[1]

I believe that we need to go beyond informal understandings of evolution, and what biologists explicitly say about it, to look at how biologists use models and statistical methods in successful empirical research, and at the roles of biological concepts in evolutionary theorizing. When we do, I argue, it turns out that natural selection and other evolutionary processes are best understood as taking place in natural, stochastic "machines" that, while realized in particular cases by actual, particular organisms and particular environmental properties, have a character that goes beyond those particular realizations.[2] This view undermines and sidesteps long-standing debates in philosophy of biology. I also argue that evolution can be viewed as produced by causal processes that occur simultaneously in multiple, overlapping systems picked out—but not constituted—by scientists' choices.

Although I've presented bits of this picture in individual papers, I believe the full picture is complex and difficult to appreciate without seeing the pieces together. The resulting argument for a view about the metaphysics[3] of evolutionary processes is both philosophy of biology and applied philosophy of probability in the context of a particular science. In my view, the kind of complexity that evolving populations exhibit, and the rich and diverse tradition of research in evolutionary biology, makes evolutionary biology the paradigmatic higher-level science of complex systems, with lessons for both philosophy of physical sciences and philosophy of social sciences. Nevertheless, I have kept the book tightly focused on evolutionary biology, because I think that certain sorts of arguments should depend on details of scientific practice in particular disciplines. Similar arguments might be given in other sciences, but trying to establish that point would require too much of a digression. Thus, even though the focus of the book is sometimes on issues that can arise only in evolutionary biology, I hope it will nevertheless be seen as one exemplar of a way in which a sustained treatment of the metaphysics of a science can be developed. I hope that some of my arguments will inspire those interested in other sciences (myself included) to apply similar or broadly analogous ideas, suitably modified.

Introduction

Overview

Natural selection, it's thought, is responsible for some—not all—changes in biological populations. Some traits or organisms are "fitter than" others, and natural selection occurs when there are changes in the distribution of traits in populations because of fitness differences. (These claims are not uncontroversial, but they will do for the moment.) Note that in evolutionary biology, "fitness" is not necessarily identical to some sort of intuitive "fit" with or adaptation to an environment. For example, sometimes biologists consider a trait fitter than another even though it often results in earlier death if it also helps more offspring to survive and reproduce. Or consider a trait that tends to increase offspring while also leading to the eventual depletion of a population's food sources. Such a trait can be considered fitter than others, at least over the short term, even if it ultimately results in extinction of a species. Thus fitness, at least according to some biological conceptions, ultimately has to do with numbers of offspring or descendants, not necessarily survival of individuals, groups, or species.

In studying evolution in populations of, say, house sparrows, a scientist might notice that during a particular five-year period, male house sparrows with bigger, darker black chest patches had, on average, more offspring. Biologists may then observe behavior to see whether darker patches affect interactions with females or with other males. They may also study genes and physiological traits connected to chest patches to see whether it's reasonable to think that some gene or trait is passed on with greater frequency because it helps male birds with larger, darker patches to have more offspring. One might then say that the inherited characteristic is selected for, or that natural selection favors it, because it makes birds with this property fitter. This need not mean that fitter birds are guaranteed to have more offspring, since there is arguably a chance element as well—but it nevertheless seems to be about actual birds in a population in the world. Because of this, it's natural to think that fitnesses are properties of individual, particular birds.

Thus, in attempting to provide systematic accounts of what biological fitness is, philosophers of biology have often started from the idea that each *individual organism* has its own particular fitness value. Yet natural selection is usually thought to be what happens when some *traits* being fitter than others results in an increase in the frequencies of the fitter traits. The result of these two assumptions—that natural selection involves fitnesses of traits in a population, and that the fundamental sort of fitness applies to individual members of the populations—led to the view that a trait's fitness should be defined as some kind of average of the fitnesses of individual members of the population that have the trait. In summary, this view is that (a) fitness differences between traits cause evolutionary changes, (b) fitnesses are, fundamentally, properties of actual, particular organisms in populations, and (c) fitnesses of traits are averages of fitnesses of the organisms in a population during some period of time. These ideas are natural if we focus closely on the idea of evolution as taking place in actual collections of organisms in the world. The view is one that in some form or other has been endorsed by many philosophers of biology, as well as some biologists.

This view is incorrect, I'll argue. It conflicts with the roles that fitness is supposed to play in evolutionary theory, and with the ways that evolutionary biologists use fitness concepts in empirical research. The assumption that a causal kind of fitness is fundamentally a property of actual individuals has, despite understandable origins, resulted in unnecessary philosophical puzzles and years of debate that I have come to see as misguided. (The contrary view that I advocate is, however, deeply indebted to a great deal of sophisticated work by those who advocated individual causal fitness views, criticized them, or elaborated their implications for various dimensions of evolutionary processes.)

In this book, I'll argue that the fitnesses of traits that are the basis of natural selection cannot be defined in terms of fitnesses of actual members of populations in the way that philosophers of biology often claim. Actual organisms in a population are just one possible manifestation of causal and probabilistic conditions that are, in a sense, embodied by the population and its environment but that go beyond their actual, realized states. It is such an overall "population-environment system," rather than actual, particular organisms living in particular environmental conditions, that is the basis of traits' fitnesses. There are some fitness concepts that apply to individuals and that play an important role in the *science* of evolutionary biology, but these don't help to make trait fitnesses involved in natural selection itself into what they are. I suspect that these claims will sound paradoxical to many readers. However, one of the points of this book is to show that, by distinguishing different classes of fitness concepts and the roles they play in the practice of evolutionary biology, we can see that evolutionary biologists' diverse uses of fitness concepts make sense together and are consistent with the idea that fitness differences cause evolution.

I also argue that when researchers in evolutionary biology choose, for example, what competing inheritable types to focus on, what set of organisms to treat as a population, or what period of time matters for evolution, their choices implicitly specify an environment relevant to processes in a population during that period of time, and thereby help to specify what fitness, for example, consists in. That is, researchers' choices delimit an aspect of the world—a population-environment system—with respect to which facts about whether and how various evolutionary processes take place are objective. I argue that different models can be viewed as selecting different overlapping real aspects of the same part of the world, so that different choices—even partially arbitrary choices—pick out different real causal relationships. This is a kind of *pragmatic realism*: pragmatic in that it depends on scientists' questions and choices, but realist in that it takes the conclusions drawn, given those questions and choices, to be about a real world independent of us (e.g., Wimsatt 2007; Brigandt and Love 2012; Waters 2014, 2017, 2019; Glennan 2017).

The book focuses narrowly on natural selection and the role of probability in evolution. Other evolutionary factors, such as random drift, mutation, biased inheritance, genetic linkage, and developmental processes, are more or less easily accommodated by my population-environment perspective, but I chose to focus the book on natural selection to allow the sort of sustained presentation that a potentially unfamiliar perspective requires. (I do summarize my view about drift, though.) I ignore, too, debates about levels and units of selection, biological individuality, and cultural evolution. Some of what is left out will appear in articles published elsewhere, as I indicate below. Despite the constrained focus of the book, however, the view that I present is designed to have much wider significance, providing a general framework for thinking about the metaphysics of biological evolution and its relations to empirical research.

In this respect, the book bears affinities to some other books that develop general philosophical characterizations of evolutionary biology, among which the best known are Sober's *The Nature of Selection* (1984), Brandon's *Adaptation and Evolution* (1990), Godfrey-Smith's *Darwinian Populations and Natural Selection* (2009b), and Pence's *The Causal Structure of Natural Selection*

(2021). The first two played a formative role in my thinking about philosophy of biology, though it will be apparent that I have come to have deep disagreements with aspects of their approaches. I don't discuss the central perspective of Godfrey-Smith's book, that we should understand evolutionary factors such as natural selection and drift as properties that lie on continua within a space of possible kinds of evolving populations. Whatever the formal and heuristic virtues of this picture, it is too unstructured to be of use to my project, which is to understand the nature of evolutionary processes given their roles in the practice of evolutionary biology. I do draw upon other elements of Godfrey-Smith's book where they are useful. Pence's book would be well worth discussing here. Among other things, it puts a critical reading of some of my papers in a larger context. Unfortunately, my book was substantially formed when Pence's book became available to me, and incorporating commentary on Pence's book would have required significant additions to mine. I intend to discuss Pence's valuable perspective in one or more later works, however.

In parts of the book, I found it useful to frame arguments in terms of an ongoing debate between two groups of philosophers of biology (and some biologists) who have come to be known as "causalists" and "statisticalists." Roughly, causalists argue that natural selection, random drift, and other evolutionary "forces" can be considered causes of evolution. Statisticalists argue that only individual organisms within populations are causes of evolution. Natural selection and drift, on this view, are "statistical" summaries of effects of actions of individual organisms, and they can only be distinguished, if at all, relative to particular models. My view is in the end causalist, but I disagree with many prominent causalist claims, and agree with many prominent statisticalist claims. I hope that readers who are not particularly interested in causalist/statisticalist debates will have patience with my sometime focus on them, as engaging with these debates helps to clarify points that will have broader interest.

I believe that my probabilistic "population-environment" perspective and the pragmatic realist view of evolution have general value, most obviously, for philosophers of biology and biologists, as well as other philosophers of science and scientists. Answering the questions I pursue in this book will, I hope, give us a deeper understanding of the natural world. I hope that even those who have never wondered about such questions will find that my arguments and proposals help to provide a deeper and more systematic understanding of evolution and its study. And though the nature of my arguments means that I have had to focus closely on some details of evolutionary biology, the overall picture that I paint is relevant to philosophy of science in general, and a number of the arguments can be adapted to social sciences and other sciences of complex systems.

Because my focus is on contemporary evolutionary biology, I won't directly engage with a large body of important and illuminating research by philosophers, historians, and biologists on the history of evolutionary biology. However, much of contemporary evolutionary biology draws directly on work considered to be part of the Modern Synthesis—developed by a few innovators in the early and middle decades of the twentieth century—which transformed Darwinian theory by using mathematical modeling to integrate Mendelian genetics into it. Thus, readers interested primarily in the Modern Synthesis or other aspects of the history of evolutionary biology may still find the perspective I provide quite valuable. (In chapter 1, I illustrate this possibility with a discussion of a seemingly unresolved debate in recent evolutionary biology that began in the last quarter of the twentieth century but that has roots in the Modern Synthesis.)

My arguments draw upon ideas from evolutionary biology—including quite a bit more than what the average well-educated person might know about evolution—and from philosophy of science. However, I have tried not to assume a background in either field. The chapter "Background on Probability and Evolution" provides a brief introduction to relevant ideas from evolutionary biology, philosophy of biology, philosophy of probability, and some other areas of philosophy of science. I introduce additional technical material as needed in later chapters, but the "Background" chapter is intended to get one started. Note that although mathematical expressions are found here and there in the text—because they have the potential to make ideas clearer for some readers—I assume only a basic acquaintance with algebra, and I try to make my points sufficiently clear in English for those whose eyes may glaze over when encountering equations. In many cases, I don't bother with mathematical expressions at all. Readers with sufficient background will be aware of the symbolic formulations behind the words.

Some people seem to think that philosophy of science is of significance only when it can help scientists accomplish what they take themselves to be doing. I see that as an unreasonable constraint on human knowledge. Among other things, it does a disservice to scientists, whose interests in understanding the world can surely go beyond questions that they have actively tried to answer, or even that they have previously thought to ask. One of the ways philosophy of science can help science's contribution to human understanding is by developing answers to questions that scientists simply don't need to answer to do their work. And it is, in fact, fortunate that not all fruitful increases in science-related understanding affect scientific practice. For example, if evolutionary biologists had not been able to study natural selection or use statistical methods without understanding clearly, in all cases, what biological fitness is, what a population consists in, what probability is, and so on—that is, without answers to the sorts of questions philosophers of science and philosophically minded scientists have tried to address—then a great deal of useful research would never have been done. Scientists often figure out how to get important research done by working around, defining away, or ignoring conceptual, epistemological, or metaphysical problems that philosophers treat as their primary subject. Sometimes such problems do interfere with scientific progress, and philosophers have helped to sort them out in some cases (Laplane et al. 2019), but that is not the general rule. It may well turn out that my arguments have implications for scientific practice, but developing those implications is not my goal in this book. If the book manages to help to provide a clearer understanding of what evolutionary processes are, of the nature and role of probability in them, and of their relationships to decisions about empirical methods, that would be enough.

Methodology

As I hope readers will agree (eventually), organisms and their interactions with each other and with their environments are all enormously complex. As a result, it's impossible to study evolutionary processes in such a way that the evolution in any population—even in the lab—is fully understood. This means that we can only study evolutionary processes using models that simplify what is actually going on, and only with the help of statistical methods that, at best, give scientists good reasons to believe that their conclusions about particular evolving populations hold—and then only approximately.

This view is consistent with the possibility that the metaphysics of evolutionary processes is, in fact, quite systematic, at least in the sense that what happens in an evolving population does so according to definite probabilities. Call this the *systematic reality* view of evolving populations. On the other hand, it may be the evolving populations do not exhibit any truly systematic behavior. On this view, scientists do manage to construct somewhat systematic descriptions using models and statistical methods, but that merely shows that something not very systematic can be (kind of, sort of) approximated using our clever devices. This is the *raw mess* view of evolving populations.

I think the truth is somewhere between the systematic reality and raw mess views, but in most of the book, I write as if the systematic reality view is correct. This is strategic. If we start from the assumption that it's all just a mess in the world, then we should give up on the possibility of finding anything systematic. If we assume that there is systematicity to be discovered in the world, by contrast, we can try to work out what it must be like. It may turn out in the end that we have to say that the world is less systematic than we'd hoped, but we're more likely to find what systematicity there is by assuming that it exists. As I see it, the real world of evolving populations is far from pretty or elegant—it is, in some sense, a mess—but the mess still has quite a bit of structure, and it is that structure that models and statistical methods in evolutionary biology capture somewhat accurately (see Wimsatt 2007; Mitchell 2009; Potochnik 2017; Waters 2017, 2019).

Some of my arguments draw inferences from patterns of practice in evolutionary biology—from how scientists invent and use concepts, what they take as evidence, and what they take the evidence to be evidence for. Some of the examples of empirical research that I choose for my case studies may seem more complicated than necessary, but much empirical research in evolutionary biology is complicated, and it's illuminating to get a peek into real science. Moreover, a relatively abstract, relatively simple, novel theory of evolutionary processes like mine should have to confront some of the details of real science, rather than only looking at idealized versions of it. Because of the complexity of some of the examples, though, it turned out to be efficient to reuse examples to make points in different parts of the book. The case studies I discuss are nevertheless often quite representative of practices that are widespread, and I'll make it clear when unusual aspects of a study matter for my arguments. I also offer arguments that depend on established empirical evidence or on broad but defensible generalizations about the biological world, and I draw inferences from the roles that certain concepts play in evolutionary biology. I sometimes use made up, very simple hypothetical examples of evolution—"toy models." Some philosophers of biology disdain their use, but toy models can be found in textbooks and in theoretical and empirical research in evolutionary biology. Such models are useful for making theoretical points and for introducing new ideas in a simple manner. My simple models are always inspired by empirical research or general facts about evolution in the natural world, as my remarks will show.[1]

Structure of the Book

GENERAL ADVICE

I present a picture that is multifaceted but integrated. Though parts are mutually supporting, there is no completely natural order of presentation. Chapters are nevertheless designed to be read in order, except that readers with sufficient background can skip unneeded parts of the chapter "Background on Probability and Evolution," which reviews ideas from philosophy of science and evolutionary biology needed for the rest of the book. (Since this chapter comes before chapter 1, its sections are numbered 0.1, 0.2, etc.) It may be possible to read some of the other chapters out of order if one is willing to jump back to earlier sections when necessary, but everyone should read chapters 1 and 2 before later chapters.

PART I: LAYING THE FOUNDATION

This part of the book outlines core ideas that will be referenced throughout the book, criticizes current philosophical conceptions of evolutionary processes, and begins to spell out positive aspects of my view.

Chapter 1, "Population-Environment Systems," and chapter 2, "Causal Probability and Empirical Practice," introduce core ideas that will play central roles in the book:

- The view that evolutionary processes exhibit what I call *lumpy complexity*.
- The concept of a *population-environment system*, which provides a foundation for the reorientation of thinking about evolutionary processes that I advocate.
- The idea that models and statistical inference provide imperfect evidence about imperfectly specified causal patterns.
- The concept of *causal probability*, a category for certain varieties of objective probability or *chance*.
- The idea of *arguments from empirical practice*, which play an important role in some parts of the book.

Using some of the preceding points, I give an argument that evolutionary processes depend on causal probabilities. I also illustrate and clarify the concepts of population-environment system and causal probability by arguing that they help us to understand the nature of a historical and ongoing debate about how to estimate F_{ST}, a measure of genetic differences between populations first introduced by Sewall Wright in 1951.

Chapter 3, "Irrelevance of Fitness as a Causal Property of Token Organisms," introduces several roles for fitness concepts that can help to sort out problems that have plagued debates about fitness concepts among philosophers and some biologists. I then provide a series of arguments that what I call "causal token-organism fitness," or "causal token fitness," cannot play the roles in evolutionary biology that philosophers intended it to play.

Chapter 4, "Roles of Environmental Variation in Selection," provides a more detailed discussion of an aspect of chapter 3: relationships between environmental variation and fitness. This chapter begins from a discussion of causal token-organism fitness, but eventually focuses on what I call "causal organism-type fitness," or "causal type fitness." I argue that only certain kinds of environmental variation are truly relevant to causal organism-type fitness. I also explain why the search for a single core definition of fitness is misguided. This is due in part to certain benefits of linguistic flexibility for a scientific understanding of the world. This chapter also introduces further details of the concept of a population-environment system.

PART II: RECONSTRUCTING EVOLUTION AND CHANCE

Though this part of the book continues criticisms of others' views, it's largely devoted to further development of a positive view about the nature of evolutionary processes. I use the foundation provided by earlier chapters to explain how we can understand natural selection as a cause in pragmatically selected populations, how relationships between fitness concepts and empirical practices support this view, and how we might understand the source of probability in population-environment systems.

Chapter 5, "Populations in Biological Practice: Pragmatic Yet Real," looks at concepts of population in evolutionary biology. This is important because a population-environment system would typically be defined in terms of a specification of a population. I examine some previous proposals to provide a definition of "population," and I argue that existing research in evolutionary biology shows that populations can be defined in extremely flexible ways.

Chapter 6, "Real Causation in Pragmatic Population-Environment Systems," explains how we can understand natural selection as a cause of evolution even though the populations in which it takes place are defined in the flexible ways indicated in chapter 5. I argue that the apparent conflicts between causal claims that result from defining multiple populations over the same individuals are not, in fact, conflicts. This chapter also includes a brief discussion of what random drift is, on my view.

Chapter 7, "Fitness Concepts in Measurement and Modeling," is where I argue for the primary roles in empirical research of three of the fitness categories defined in chapter 3. I argue that causal token-organism fitness plays no role in such research. I discuss ways that a kind of token fitness sometimes plays a role in theoretical modeling, but I argue that the token fitness concepts involved are very different from the kind that some philosophers have promoted.

Chapters 8 and 9, "Chance in Population-Environment Systems" and "The Input Measure Problem for MM-CCS Chance," explore what I take to be reasonable proposals about the nature of the causal probabilities realized in population-environment systems. Among other things, I argue that these probabilities can't be single-case propensities, and I discuss the idea that population-environment probabilities might be some variant of what I now call "measure-map complex causal system" probabilities, introduced separately by Jacob Rosenthal, Michael Strevens, Wayne Myrvold, and me.

The concluding chapter summarizes the picture developed in the book and invites responses to it.

I can explain now that this book concerns "evolution and the machinery of chance" in at least two senses: First, I present population-environment systems as complex machinelike systems with chances—specifically, causal probabilities—of various possible outcomes. In this sense, chances are embedded in and filtered through the complex machinery that gives rise to evolutionary outcomes, whatever the source of those chances may be. Most of the chapters of this book are devoted to exploring details of this strategy. Second, chapters 8 and 9 then investigate the possibility that the complexity of population-environment system machinery is itself the source of the chances affecting outcomes relevant to evolution.

GOING FURTHER

There are several related papers that will, I hope, be published near the time that this book comes out. These include work on genetic drift (Abrams Drift MS), infinite-population concepts and modeling (Abrams InfPops MS), a critique of some particular fitness concepts (Abrams Fitness MS), a solution to a problem that may face some views about probability in population-environment systems (Abrams LongRun MS), and a discussion of pseudorandom number generators (PRNGs) as generators of chance (Abrams PRNG MS). A paper in which I argue that evolution depends on what is known as objective imprecise probability has already been published (Abrams 2019).

0

Background on Probability and Evolution

0.1 Introduction

This chapter provides background context that, I hope, will allow every interested reader to get enough of a handle on material in later chapters that additional pointers and explanation provided there will be useful. I provide brief introductions to philosophical, scientific, and mathematical ideas about the nature of probability (§0.2), and introductions to ideas about natural selection, biological fitness, random drift, and other evolutionary processes (§0.3). I define potentially unfamiliar terminology either in the main text or in endnotes. Some readers can skip this chapter, but I encourage everyone to at least skim it to find sections or paragraphs that may be helpful—perhaps only because I frame a familiar idea in a useful way. Other chapters will include occasional references to material in this chapter to help those who find they've missed something valuable. Those who find my treatment of topics in this chapter disappointingly insubstantial should not worry too much; substance has been left for later chapters.

0.2 Probability and Philosophy

I assume that readers have at least some intuitive familiarity with probability concepts. The main point of the next section is to emphasize the existence of an abstract, mathematical notion of probability. This will allow me to put probability concepts of a more applied or philosophical nature in relation to the abstract concepts in subsequent sections.

0.2.1 MATHEMATICAL PROBABILITY

From the point of view of mathematics, probabilities are values of any mathematical function P( ) that satisfies a few rules concerning subsets of a set Ω. The subsets must be "closed under finite unions," which means that if P( ) applies to two sets A and B—that is, P(A) and P(B) have values—then P( ) also applies to the union $A \cup B$ of those sets—that is, P($A \cup B$) has a value. Sometimes it's also required that these sets be "closed under countably infinite unions," so that if we have an infinite sequence of sets $A_1, A_2 \ldots$ (perhaps getting smaller), the probability of their union P($\cup_{i=1}^{\infty} A_i$) is also defined.[1] Traditionally, the subsets are called "outcomes," or "events," because they usually represent sets of possible things that could happen in the world.[2]

Here are the rules, or axioms, that a probability function P( ) must satisfy:

1. The function P(A) assigns a nonnegative (real) number to every allowed set A.
2. Additivity:
    a. (Weaker version) The probability of either A or B occurring, where A and B are mutually exclusive, is the sum of A's and B's probabilities: P($A \cup B$) = P(A) + P(B) for any two subsets A and B that have no elements in common.
    b. (Stronger version) P($\cup_{j=1}^{\infty} A_j$) = $\sum_{j=1}^{\infty} P(A_j)$ for A_js that have no elements in common. That is, the probability of an infinite union of sets sharing no elements is equal to the sum of the probabilities for each of the sets. This version is usually paired with the requirement that the set of subsets is closed under infinite unions.
3. The entire set Ω has probability 1: P(Ω) = 1.
4. The probability P(ø) of the empty set ø is zero.

The rest of this section reviews common probability terms and concepts that will occasionally be important later. Some readers may want to skip to the beginning of section 0.2.2.

A standard mathematical definition of *conditional probability* P($A|B$) is the probability of A and B divided by the probability of B:[3]

$$P(A|B) = P(A \cap B)/P(B).$$

Informally, this is the probability of A within all situations in which B occurs.

Probabilities of two outcomes A and B are called *independent* when the probability of A given B is equal to the probability of A, simpliciter:

$$P(A|B) = P(A).$$

This means, informally, that whether one is in a B situation doesn't make a difference to the probability of A. Another definition of independence, which can be derived from the preceding equations, is that A and B are independent if and only if the probability of their conjunction is equal to the product of their probabilities:

$$P(A \cap B) = P(A) \times P(B).$$

Similarly, A, B, and C are jointly independent if and only if

$$P(A \cap B \cap C) = P(A) \times P(B) \times P(C).$$

It's important to realize that if A and B are independent, and B and C are independent, that doesn't necessarily mean that A and C are independent, nor that A, B, and C are collectively independent. That point generalizes to more outcomes, of course.

When outcomes are numerical values, or when we are interested in some mathematical function of outcomes whose values are numerical, we can treat those numerical values as the result of a *random variable*—a technical term for a function whose values have probabilities.[4] For example, the number of heads in ten tosses of a fair coin is a random variable, each of whose possible values 0, 1, 2, . . . 10 has a probability.

A *distribution function* specifies the probabilities of values, or ranges of values, of a random variable. A *density function* for continuously varying values of a random variable assigns a number for each value in such a way that the integral over a continuous set of values is the probability of those values (Grimmett and Stirzaker 2001).[5] In later chapters, I often use "distribution" in a broad sense that can refer to either a density function or a distribution function proper, and I sometimes speak of the distribution of actual outcomes to describe a pattern of frequencies over outcomes. Context should make the meaning clear.

To *partition a set* such as Ω is to divide it up into mutually exclusive subsets—*members of a partition*—so that every element is in exactly one subset. Thus, for example, the additivity requirement above implies that the sum of the probabilities of members of a partition of Ω will be equal to the probability of Ω (i.e., 1).

0.2.2 BEYOND MATHEMATICAL PROBABILITY*

Consider the question, What is the probability of the portion of the page that is above the beginning of this printed sentence? That question should sound odd: in what sense do parts of a page have probabilities? From a purely mathematical point of view, though, there need be nothing wrong with the

---

* This section contains brief excerpts from Abrams (2012d): Marshall Abrams, "Mechanistic Social Probability: How Individual Choices and Varying Circumstances Produce Stable Social Patterns," in *Oxford Handbook of Philosophy of Social Science*, ed. Harold Kincaid, 184–226 (Oxford University Press, 2012). Reprinted by permission.

question. If we divide up (partition) the page into small squares, the proportion of the area of the page in each square, or in unions of squares, will satisfy the preceding axioms. Proportion of the area of the page can thus be seen as an entirely legitimate kind of probability—in the mathematical sense. The reason that the question sounds odd is that in the way we normally intend "probability," subsets of pages are just not the sorts of things that count as having probabilities. Thus it's not enough to satisfy the axioms to count as probability in the everyday—or scientific—sense. More is required. At the very least, probability in science must

1. concern situations, events, states, and so on that could occur in the world;
2. satisfy mathematical axioms of probability; and
3. satisfy other criteria that capture roles of probability in scientific contexts.

There is disagreement about what the criteria mentioned in (3) should include. However, probability plays different roles in different contexts, so it seems implausible that a single set of criteria will be universally appropriate. At the very least, we can distinguish probabilities that in some sense capture facts about degrees of confidence, and probabilities that capture facts about relationships in the world. The former are called *Bayesian, subjective,* or *epistemic probabilities*; the latter, *objective probabilities* or *chances*.

For example, suppose that we are tossing a die that is loaded: the material that it's made of is not uniformly dense, and it is weighted more heavily opposite one side. If you don't know how the die is loaded, your degree of confidence in each of the six outcomes might be the same (subjective probability), yet it seems intuitively clear that one of the outcomes—the one opposite the extra weight—has a greater probability. The latter is an objective probability, because it depends solely on facts about the world, and it can be independent of our degrees of confidence. I'll always use "chance" to mean objective probability, as opposed to Bayesian subjective or epistemic probability, though the subjective probability won't play a large role in this book.[6]

There is a further question about the die, however. The die is just a cube with labeled sides, and tossing it is a physical activity. What is it about tossing a die that makes the label that is facing up when the die stops moving into an outcome with a probability? What is it about die tossing that gives rise to chances? We can ask similar questions about other games of chance. More importantly, if there are chances of outcomes that matter in evolutionary processes—as I'll argue there are—then we can ask what it is about those outcomes, and how they are embedded in biological or other physical processes, that gives them chances. Traditionally we say that this is to ask which "interpretation of probability" defines the probabilities in question. This formulation is, in my view, a bit misleading, because interpreting is an activity that humans engage in. What we want to know is, What is it about the world that makes it the case that there really are probabilities in a given situation?

0.2.3 INTERPRETATIONS OF PROBABILITY*

I'll mention, very briefly, some well-known interpretations of probability. My goal is not to do justice to them but merely to illustrate the concept of an interpretation of probability, and to use some obvious drawbacks of these interpretations to suggest that something else may be needed for understanding evolution. That will set up some discussions later in the book.[7] One of the interpretations will also turn out to be important for a popular philosophical theory of biological fitness. Before we get started, though, note that "frequency" can refer either to *absolute frequency* (700, in 1000 trials) or *relative frequency* (700/1000 = 0.7, or 70%). Context will usually make it clear which I intend, but sometimes I'll make the distinction explicit.

Bayesian interpretations are those that define probabilities as degrees of belief (or "credences") or define them in terms of epistemic relationships. Bayesian probabilities are the starting point for an important set of approaches in statistics (e.g., Gelman et al. 2013), and they may be important for understanding scientific inference in general (e.g., Earman 1992; Howson and Urbach 1993; Sprenger and Hartmann 2019).[8] However, it seems that Bayesian probabilities can only reflect and track frequencies in the world—not cause or explain them. The role that probability plays in evolution, I'll argue later, requires that probabilities have something like a causal relationship to frequencies.

Next consider a simple *finite frequency* interpretation[9] of probability, according to which probabilities in science are identified with relative frequencies of outcomes—for example, days on which a particular purple finch finds berries—in an actual set[10] of occurrences—for example, days in the life of the house finch. Among other problems, this interpretation implies that *improbable* combinations of outcomes cannot come to have a high frequency "by chance"—if they did, they would by definition not be improbable (Hájek 1996). That is, a finite frequency interpretation cannot make a distinction between fundamentally chancy regularities and accidental regularities. For example, if a die is tossed ten times, the probability of rolling a six is simply the

---

* This section contains brief excerpts from Abrams (2012d). Reprinted by permission.

number of times that six comes up divided by ten. Whether the die is loaded makes no difference.

Long-run frequency interpretations define probability as frequency in large, counterfactual sets of events, or as the limit of frequencies in a counterfactual infinite sequence. Long-run frequency interpretations face deep problems that I won't review (Hájek 1996, 2009). I will point out that neither long-run frequency nor finite frequency interpretations can explain frequencies of actual events: the events whose frequencies define these interpretations include those events whose frequencies are to be explained. In the case of a simple finite frequency theory, a probability is the frequency in an actual set of events; it therefore can't explain that same frequency in that same set of events. In the case of a long-run frequency theory, a set of actual events is supposed to be a subset of the full counterfactual sequence of events that defines a probability. Putting aside problems about the explanatory power of such merely counterfactual frequencies, a frequency in a set or sequence of events will not explain a frequency in a proper subset of those events without additional substantive assumptions (e.g., assumptions about processes that sample from the larger set/sequence).

Best system interpretations of probability (Lewis 1980, 1994; Loewer 2001, 2004, 2020; Hoefer 2007, 2019) are currently popular; they do a good job of addressing certain philosophical problems. These theories are difficult to summarize quickly, though. One core idea of the best system approach is, very roughly, that an objective "chance" exists whenever assuming its existence would play a simplifying role in an ideal, ultimate set of scientific theories— those that succeed well in summarizing all actual events in the universe, past, present, and future. What best system theories don't do well, as I see it, is allow probabilities to explain or cause frequencies, because best system chance probabilities are defined in terms of whatever states the actual world has— which include the frequencies of outcomes.[11]

Propensities are postulated probabilistic dispositions.[12] According to some common philosophical theories concerning causal properties, a lump of salt sitting in a salt shaker has deterministic *disposition* to dissolve when placed in water at room temperature, even if it is never placed in water. Often the disposition is identified with those physical properties of the salt that would interact with water to result in the salt dissolving. Propensity is a proposed extension of this idea. For example, some authors claim that an evenly weighted coin has an indeterministic disposition of strength 1/2 to produce the outcome heads in a particular case in which it's tossed in the usual way. This is what's known as a *single-case propensity*, which exists if there are indeterministic dispositions for a situation to produce any of several mutually exclusive outcomes at a particular time and place. One can then mathematically derive propensities for outcomes in sets of trials. A *long-run propensity* (Gillies 2000) instead involves a disposition to produce frequencies in large numbers of trials, where this disposition is not derived from single-case propensities. (An alternative kind of conception of single-case propensities due to Mellor [1971] and Suárez [2013, 2017a, 2017b, 2020] treats propensities as nonprobabilistic dispositions to produce probability distributions. The distinction between this approach and the one in which propensities themselves realize probabilities won't make a difference to my discussion. The Mellor/Suárez approach is complex, though, and it would complicate later discussions of propensity. The Mellor/Suárez view is important, however, so I regret that I can't discuss it further. See also note 1 to chapter 7.) An important point is that since propensities are supposed to be dispositions, and dispositions are supposed to involve causal properties, propensities would have the potential to bear causal relationships to outcomes and frequencies. On the other hand, some authors argue that propensity theories are too vague and unmotivated to be explanatory, and there are other objections as well (Eagle 2004; Berkovitz 2015; see also §2.2.2). In any event, the idea of propensity has played a significant role in philosophical thinking about biological fitness, and I discuss it further below and in chapters 3 and 8.

Note that we can generalize the single-case/long-run distinction beyond propensity. A single-case chance, or single-case objective probability, is an objective probability for the outcome of a single "trial" or instance of a "chance setup" (see §1.3). A long-run chance is also an objective probability, but it concerns patterns of outcomes only in large numbers of trials. My discussion of interpretations of probability continues in chapters 1, 2, 3, 8, and 9.

0.3 Evolutionary Biology and Philosophy

This section sketches some ideas from evolutionary biology and philosophy of biology needed as context for later chapters. My discussion of evolutionary biology is informed by distinct but interrelated theoretical fields (e.g., population genetics, quantitative genetics, statistical genetics) that are used to model evolutionary processes and that often provide the basis for statistical inference in empirical research.

0.3.1 "TYPE" AND "TOKEN"

A distinction between particular individual organisms—*token* organisms—and organisms' traits—organism *types*—will play a central role in this book.

This terminology comes from philosophy, but everyone is familiar with the general idea. For example, the three-letter string "AAA" contains three tokens of a single letter type, *A*. Each of the three "A" tokens has the type *A*, or the property of being an *A*. Each also has the property of being a letter—they are instances of the letter type—as well as being instances of other types that have to do with shape or linguistic role, for example. I need the type/token distinction to distinguish unambiguously between particular, individual, token organisms, on one hand, and their traits (i.e., types) on the other. Biologists' terms for token organisms, "organism" and "individual," are ambiguous in ways that could cause confusion in my discussion.[13] For example, two humans and three cats live in my house. Which *organisms* are in the house? Humans and cats (types), or Marshall, Brenda, Wally, Harriet, and Elliott (tokens)? "Individual" is ambiguous in a different way. Michod (1999, 9) wrote, "Fitness is often defined as the expected reproductive success of a type.... I refer to this notion of fitness as individual fitness." Michod's point was that some models of evolution allow properties (types) of *groups* of organisms to have fitnesses. Michod was defining a term for fitnesses of properties of *organisms* rather than groups. In the rest of the book, I'll use "type" and "property" interchangeably. Note that organisms' traits are types, but some properties of organisms, such as being something that is currently perching on a branch of tree, are not naturally referred to as traits.

0.3.2 FITNESS

Fitness concepts play an important role in evolutionary biology. They appear both in informal conceptions of biological processes and in particular biological models. It's often said, at least colloquially, that natural selection is the "survival of the fittest."[14] What does it mean to be fittest, and what is it that is supposed to survive? For example, suppose we say that survival is not just survival per se but that it also involves producing offspring. And suppose we say that for one (token) organism to be fitter than another is for it to actually have more offspring that survive and reproduce. Then it looks like the survival of the fittest is just the survival of those that best survive. Yet natural selection is also thought to explain many things: why organisms seem adapted to their environments, the distribution of species, the fossil record (Darwin [1859] 1964), even why humans have backaches (Nesse and Williams 1994). How can a simplistic tautology—"Those that best survive are those that best survive"—explain anything? How can fitness differences in this sense help to cause evolution? We have to look a bit deeper.

Darwin gave the following well-known characterization of natural selection in the first edition of *On the Origin of Species*:

> Owing to this struggle for life, any variation, however slight and from whatever cause proceeding, if it be *in any degree profitable* to an individual of any species, in its infinitely complex relations to other organic beings and to external nature, will *tend to* the preservation of that individual, and will generally be inherited by its offspring. The offspring, also, will thus have *a better chance* of surviving, for, of the many individuals of any species which are periodically born, but a small number can survive. I have called this principle, by which each slight variation, if useful, is preserved, by the term of Natural Selection. (Darwin [1859] 1964, 61; emphasis added)

Notice Darwin's suggestion that some "profitable" variations will "tend" to facilitate survival, and that some offspring will have "a better chance" of surviving. This suggests that differences in fitness might have to do with tendencies and chances, rather than actual numbers of offspring. Note that evolution is a process that can take place over many generations, and over long periods of time. Often no organism will survive for the entire period over which natural selection is thought to act. It therefore seems that natural selection should be able to act on types that can appear repeatedly over different periods. It must be possible for these types to be inherited. They must be genes, genotypes, or phenotypes.[15] Finally, note that a genotype or phenotype A having greater fitness than another B in the same population is plausibly taken to be a necessary condition for natural selection of A over B to take place, and relationships between fitnesses seem to be an essential part of natural selection. (I'll ignore important cases of natural selection involving equal fitnesses.)

It's worth noting that there are quite a few names for the diverse variety of different fitness concepts in use in evolutionary biology, including "selection coefficient," "lifetime reproductive success," "adaptive value," "reproductive value," "net reproductive rate," "intrinsic growth rate," "density-independent growth rate," "Malthusian parameter," "viability," "number of offspring," "selection differential," "selection gradient," "invasion fitness," and others (e.g., Stearns 1976, 1992; Lenski et al. 1991; de Jong 1994; Michod 1999; de Valpine 2000; T. F. Cooper, Rozen, and Lenski 2003; Elena and Lenski 2003; Ewens 2004; Gillespie 2004; Rice 2004; Metz 2008; Shaw et al. 2008; G. Wagner 2010; Reydon 2021). Although fitness is often directly associated with numbers of offspring or descendants, fitnesses of plants are sometimes measured using the total mass rather than number of offspring (e.g., Dong, van Kleunen, and Yu 2018; Hamann, Weis, and Franks 2018), and fitness is sometimes defined in terms of physiological properties. For example, Morris, Lenski, and Zinser (2012) define the fitness benefit of certain mutations in terms of energy savings or reduced resource usage. Note that fitnesses generally depend on the environment in which the organism lives. For example, if survival is more probable with fur that camouflages an animal, the degree to which it is camouflaged will depend on its surroundings as well as its fur coloring.

The various fitness concepts are not interchangeable, but the ones above are all directed, in some sense, toward capturing properties relevant to representation of descendants or traits at later times. Discussing their similarities and differences would require a very long chapter, but I'll be able to generalize over them in later chapters, and I'll use "fitness" for all such fitness concepts. Some fitness concepts have a different focus and are primarily intended to capture the intuitive idea of "fit" with an environment. Although some sort of fit can be relevant to future representation of descendants or traits, this is not in itself part of my focus. Thus, I won't discuss the notion of "adaptation" that is the primary focus of Reeve and Sherman (1993). However, I do include under my term "fitness" those concepts of "adaptedness" (Brandon 1978, 1990; Byerly and Michod 1991) that primarily concern future representation in a population.[16]

An important tradition in philosophy of biology began about forty years ago, when Robert Brandon (1978) and Susan Mills and John Beatty (1979) argued that fitness is rooted in objective probabilities—in particular propensities (§0.2.3)—rather than actual numbers of offspring. Brandon and Mills and Beatty first defined fitness for token organisms in terms of the expectation E, or expected number of offspring, a probability-weighted average of the number O_i of offspring for a token organism:[17]

$$(0.1) \qquad \text{fitness of organism } i = E(O_i) = \sum_{k=0}^{\infty} k\, P(O_i = k)$$

$$= 0 \times P(O_i = 0) + 1 \times P(O_i = 1) + 2 \times P(O_i = 2) + \cdots.$$

Here $O_i = k$ means that organism i has k offspring. This definition of fitness makes it the sum, for all possible offspring counts k, of the products of k and the probability of an organism having k offspring. (The probabilities will be zero for large k, since litter sizes and lifetimes are bounded, so this is in effect a finite sum.)

The general idea of defining fitness as expected number of offspring was not invented by philosophers; biologists often define fitness concepts using expectations of various sorts for numbers of offspring (see chapter 7). However, Brandon and Mills and Beatty made it explicit that, initially, (a) the number of offspring is for each particular, token organism, and (b) the probability in terms of which the expectation is to be calculated is a propensity in the sense described above in section 0.2.3.[18] An advantage of this way of understanding fitness in terms of propensity is that it seems to make sense of fitness's apparent causal and explanatory force (e.g., Mills and Beatty 1979; Millstein 2016). Just as salt's disposition to dissolve in water bears a causal relationship to a lump of salt dissolving, so differences between the dispositions of token organisms to produce various numbers of offspring bear a causal relationship to some traits or genotypes becoming more or less common in subsequent generations. Following Mills and Beatty's suggestion, the new definition of fitness in terms of propensity, as well as some variations on it, has become known as the "propensity interpretation of fitness" (PIF).

Mills and Beatty (1979) and Sober (1984) also defined a concept of fitness for trait in roughly this way:

> The fitness of a trait in a population during a (short) period of time is the average of the fitnesses of those organisms in the population with that trait during that period.[19]

This view is not always made explicit by PIF advocates—often their focus is primarily on fitnesses of token organisms—but variants of it are still defended (e.g., Pence and Ramsey 2013; Ramsey 2013a, 2013b; Sober 2013). I criticize this simple averaging view in chapter 3; the view of trait fitness that I develop in subsequent chapters implies that recent more subtle variants are on the wrong track too.

PIF advocates (Beatty and Finsen 1989; Brandon 1990; Sober 2001) and others questioned the definition of fitness as expected number of offspring because of Gillespie's arguments (1973, 1974, 1975, 1977) that a type with a lower expected number of offspring than another can have a greater probability of reproductive success over the course of many generations. Gillespie's insights expanded the theoretical basis for evolutionary biology (Slatkin 1974; Philippi and Seger 1989) and supported new insights in empirical studies (e.g., Gremer, Crone, and Lesica 2012). Gillespie's work also led to a broad range of responses by philosophers, both as inspiration for improved versions of the PIF (e.g., Beatty and Finsen 1989; Brandon 1990; Sober 2001; Abrams 2009b; Pence and Ramsey 2013; Takacs and Bourrat, 2022) and as the basis of criticisms of it (e.g., Ariew and Ernst 2009; Walsh 2010, 2015b). Most of the philosophical debate stemming from Gillespie's work draws upon the first pages of Gillespie (1977), a short discussion inspired by his earlier papers. There Gillespie argued that in some simple cases, if we want to predict which of two competing traits will outreproduce the other, we must use variance as well as expectation of number of offspring to define fitness. In one sort of simple case, if we define the fitness w of a trait in terms of the expectation $E(O)$ and variance $\text{var}(O)$ of number of offspring like this, as Gillespie suggests,[20]

$$w = E(O) - \frac{1}{2}\text{var}(O),$$

then the fitter trait will be the one that is likely to increase in frequency over the long term. In certain other (also simple) cases, a better prediction uses this formula:

$$w = E(O) - \frac{1}{N}\text{var}(O),$$

where N is the population size. Simple illustrations of these points can be found in Beatty and Finsen (1989), Brandon (1990), Sober (2001), and Abrams (2009b). Philosophers have sometimes focused on variance as *the* mathematical property other than expectation that is relevant to fitness (e.g., Brandon 1990; Sober 2001; Walsh 2007; Ramsey 2013a), even though Gillespie's claims about variance only concerned specific sorts of cases with simplifying assumptions that are not always reasonable.[21] I discuss some of the issues raised by Gillespie's work in chapter 4 and others in Abrams (2009b), as well as in unpublished work (Abrams Fitness MS, Drift MS).

Bouchard (2008, 2011) championed the idea that fitness concepts should treat survival as more than a step on the way to producing offspring. Rather, the length of time that an organism survives can directly contribute to something that deserves the name of fitness, in part because organisms that persist for a long time may continue to contribute to a population. Bouchard's view is also designed to address difficult questions about symbiosis between different organisms. These issues are important, and I think Bouchard's approach is worth taking seriously. I believe that the population-environment system conception that I develop below can easily accommodate Bouchard's approach, but I won't argue for this point, because I want to focus on ideas about fitness and natural selection that are more central to contemporary evolutionary theory.

0.3.3 A WRIGHT-FISHER MODEL OF NATURAL SELECTION

As we'll see, evolutionary processes are often studied using models that ignore details of those processes, representing them in an intentionally simplified way. An enormous variety of such models is used in evolutionary biology. In this section and the next, I introduce one variant of a well-known mathematical model of natural selection, the Wright-Fisher model, both as background for later sections in this chapter and as a simple illustration of ideas that will be used throughout the book. My aim is to give a reader unfamiliar with Wright-Fisher models some insight into their character.

Suppose we have a population of fixed size N containing seasonally reproducing, sexually reproducing, diploid organisms with alternative alleles A and B at a locus that we'll track.[22] There are thus three genotypes in the population: AA, AB, and BB. Each individual has exactly one of those genotypes. We'll assume that mating is random and that the population is in (and will remain in) what's known as Hardy-Weinberg equilibrium. This means that the frequency of each genotype is the product of the frequencies of the allele types. For example, the frequency of the AA genotype is the square of the frequency of A in the population.

We want to calculate probabilities of changes in the frequencies of the A and B alleles from one generation to the next. Suppose that each individual in the parent generation contributes a large number of gametes to a pool from which N individuals are selected, and suppose that each allele is selected from the pool of gametes independently.[23] Strictly speaking, this should use what is known as *sampling without replacement*: when one gamete is "chosen" or sampled for the offspring generation, there is one fewer gamete that can be selected. However, with a large number of gametes in the pool, it's reasonable to model this as a case in which the number of gametes does not shrink with each allele sampled. Thus the model below will use sampling with replacement.

Let i represent the number of A alleles present in one generation and j the number A alleles in the next generation. Notice that since there are N organisms in the population, we can also think of the population as containing $2N$ alleles. Suppose that the AA, AB, and BB genotypes differ in fitness, with fitnesses w_{AA}, w_{AB}, and w_{BB}, respectively. In our simple model, we can think of these fitnesses as numbers of gametes produced by each genotype. These are *absolute fitnesses*, because the three fitnesses do not sum to 1. We have to weight the chance of each genotype contributing to the next generation by its *relative fitness*, which is the absolute fitness divided by the average fitness of the entire population at a given time. Incorporating this idea, it can be shown that the probability p_i of an A allele contributing to next generation when the frequency of A alleles is i is (Ewens 2004, eqs. 1.25, 1.59)

$$(0.2) \qquad p_i = \frac{w_{AA}i^2 + w_{AB}i(2N-i)}{w_{AA}i^2 + 2w_{AB}i(2N-i) + w_{BB}(2N-i)^2}.$$

This will look unfamiliar to some readers, but we can break it down into its parts.

The numerator of equation (0.2): This calculates a fitness for the A allele as a weighted average of fitnesses w_{AA} and w_{AB} of AA and AB individuals in the current generation, respectively. The AA and AB individuals are those that have the ability to contribute an A allele to the next generation. The average in the numerator is weighted by the frequency, i^2, of AA individuals and the frequency, $i(2N - i)$, of AB individuals: the AA genotype's frequency i^2 is simply the frequency of A in the population times the frequency of A; we assumed that mating is random, so the frequency of combinations of A with A is equal to i^2. Similar reasoning justifies the frequency of $i(2N - i)$ for the AB genotype, since the frequency of the B allele is the total number of alleles, $2N$, minus the number of A alleles.

The denominator of equation (0.2): This adds the quantity in the numerator to the analogous calculation for the fitness of the B allele. The total in the denominator is thus the total fitness of all of the individuals in the population.

All of equation (0.2): Thus the entire ratio on the right-hand side of (0.2) is the proportion of all fitness that belongs to the A alleles. It's the average fitness of A alleles, or the fitness of any one randomly selected A allele. We will be treating fitness as determining the probability of being selected for the next generation, so this ratio also gives the probability of an A allele being chosen. Note that the probability q_i of the other alternative—that is, a B allele being chosen—is

$$q_i = 1 - p_i.$$

Using these definitions of the probabilities p_i and q_i of an A or B allele being copied into the next generation, the *transition probability P_{ij}* for an *entire* population with j A alleles to be produced from a population with i A alleles is (Ewens 2004, eq. 1.59):

$$(0.3) \qquad\qquad P_{ij} = \binom{2N}{j} p_i^j q_i^{2N-j}.$$

We can unpack this equation too. Because we're assuming that the alleles are sampled independently, the probability of selecting a sequence of N alleles in which an A appears j times and a B appears $(2N - j)$ times is the product of the probabilities of selecting each allele—that is, p_i for the A alleles and q_i for the B alleles. The probability of such a sequence is thus $p_i^j q_i^{2N-j}$. However, since the order in which the alleles are chosen doesn't matter, we have to multiply this quantity by the number of sequences containing exactly j A alleles and $2N - j$ B alleles. The number of such sequences is the *combination* $\binom{2N}{j}$, which can be defined as $\frac{2N!}{j!\,(2N-j)!}$, and which gives the number of ways that exactly j items can be selected from a set of $2N$ items without regard to the order of the selected items. Here expressions like $k!$, or k factorial, mean, for $k!$, $k \times (k - 1) \times \cdots \times 2 \times 1$. (The reasoning that justifies the definition of $\binom{N}{j}$ is not very complicated, but it would require too much of digression here.) That ends the mathematical explication of the Wright-Fisher model with natural selection. Let's look at some of its implications.

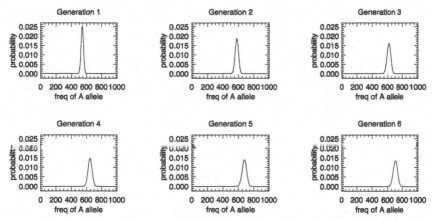

FIGURE 0.1 Probabilities of frequencies of allele A across six generations, using a diploid Wright-Fisher model with natural selection. Population size is 500, for a total of 1000 alleles. Initial frequency of A was 500. Genotype fitnesses are $w_{AA} = 1.0$, $w_{AB} = 0.95$, $w_{BB} = 0.7$.

In figure 0.1, we see an illustration of consequences of equation (0.3). This figure displays probabilities of absolute frequencies of the A allele over six generations in a population of size $N = 500$ ($2N = 1000$ alleles), with the A allele starting at a frequency of 500 in generation zero, and with fitnesses of the three genotypes as

(0.4) $$w_{AA} = 1.0, \, w_{AB} = 0.95, \, w_{BB} = 0.7.$$

Thus each curve represents the probabilities that the population will have various possible frequencies of A in that generation.[24]

This is a case in which A is fitter than B in each generation. That's why the probability curve shifts toward higher frequencies for A in each generation. The plots illustrate the idea that when natural selection tends to increase the frequencies of some types and decrease the frequencies of others, it does so only in a probabilistic manner. According to this Wright-Fisher model, no particular change in frequency is guaranteed in any generation, even though some changes are more probable than others. It is the probabilities of higher frequencies that increase over time.

0.3.4 RANDOM DRIFT

Random genetic drift, random drift, genetic drift, and simply *drift* are terms used to refer to ways in which the genetic or phenotypic composition of a population can change solely due to chance, rather than, say, natural selection. Terms referring to drift are used in different, related ways in different contexts (Plu-tynski 2007). For example, in some models, drift depends solely on the chancy process by which gametes end up forming zygotes, but in other contexts drift is treated as if it depended on other chance processes in nature. Biologists often refer to drift as a "force" affecting the composition of a population, as if it were distinct from other "forces" such as natural selection and mutation (e.g., Gillespie 2004). I see no reason to view most such talk as anything more than use of an informal metaphor, but some biologists, such as Roughgarden (1979), can be read as taking the term to mean more. Among philosophers, Sober (1984) initiated the idea of treating the force terminology seriously, describing drift, selection, and other "forces" in analogy with Newtonian forces as causal factors affecting the future makeup of a population of organisms (see also Stephens 2004, 2010; Pence 2017). Around the same time, John Beatty (1984) raised questions about whether it is possible to distinguish drift from selection. Since then, there have been ongoing debates between philosophers of biology about the nature of, and relationships between, natural selection and random drift. I summarize my view about drift and selection in section 6.5, and I provide a longer discussion in works mentioned there. Here I'll just introduce some basic ideas about drift from evolutionary biology, in order to provide context for my discussion of natural selection.

Most biological uses of "drift" treat it as a causal factor whose strength varies inversely with population size. That is, other things being equal, as population size decreases, it becomes more probable that what happens in

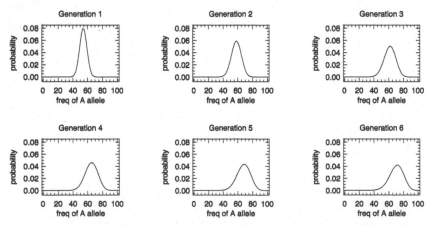

FIGURE 0.2 Plots for a model like the one in figure 0.1 but with population size 50 (100 alleles) and initial frequency 50. Genotype fitnesses are as before. Note scales differ from those in figure 0.1.

the evolution of a population will occur purely by chance, without regard to other forces of evolution such as natural selection. On the other hand, the term "drift" is also often applied to the *product* or result of the *process* of drift; see Gildenhuys (2009) for illustrations (also see Hodge 1987; Millstein 2002; Stephens 2004; Brandon 2005). For example, figure 0.2 plots results of a Wright-Fisher model like the one above, but with smaller population size. It shows there is a reasonable chance that the relative frequency of the A allele will remain at 50% for all six generations plotted, even though there is selection for the A allele. If this happened in a real population—if the frequency of A remained the same despite it being fitter—one might say the relative frequency of A remaining at 50% *was* drift.

Wright-Fisher models play a central role in theorizing and empirical research about drift, and they can be used to illustrate the idea of drift as a causal factor. Compare figures 0.1 and 0.2. Figure 0.2 is based on the same two equations, (0.2) and (0.3), as figure 0.1, and it uses the genotype fitnesses specified in equation (0.4). The initial relative frequency of the A allele is again set at 50% of the population size. The only difference between the models is population size. In figure 0.2, the population has size $N = 50$ with $2N = 100$ alleles, rather than size $N = 500$ with $2N = 1000$ alleles as in figure 0.1. Natural selection still takes place in the second model: the largest probabilities for frequencies of the A allele gradually shift toward higher frequencies. However, because the population size is smaller, drift is greater. That is, in the small-population model, the probabilities that the A allele will exhibit various relative frequencies are more spread out than in the large-population model. For example, in generation 6, the modal (most extreme) probability is at about 70% in both models, but we can see from figure 0.2 that the small-population model assigns probabilities that are significantly larger to relative frequencies between 50% and 60%, and between 80% and 90%. This illustrates the idea that in smaller populations, there is a greater chance that the course of evolution of the population will not reflect forces such as natural selection. In larger populations, the probability distributions over possible states of a population at a given time will be narrower, so that natural selection's likely effects will be more specific. Note that if all fitnesses were equal in equation (0.2)—that is, that there was no natural selection—the equation could be shown to reduce to

$$(0.5) \qquad\qquad p_i = \frac{i}{2N}.$$

This means that the probability p_i of selecting an allele for the next generation is simply the relative frequency $i/2N$ of the allele in the population. This makes a pure drift model, in which all changes of frequency are due to chance.

The following example might provide more insight about drift. Suppose that you have an opaque bag with two red marbles and one yellow marble. You draw out a marble without looking, record its color, put it back, shake up the bag, and draw again. If you do this three times, it's most likely that you will pull out more red marbles than yellow marbles, but there's about a 26% chance that you'll draw the yellow marble more often than either of the red marbles. Assuming that the probability of choosing the yellow marble on each draw is 1/3, the probability of choosing the yellow marble exactly twice in three draws is $\binom{3}{2}\left(\frac{1}{3}\right)^2\left(\frac{2}{3}\right)^1 = 3 \times \frac{1}{9} \times \frac{2}{3} = \frac{2}{9}$, while the probability of choosing the yellow marble three times is $\binom{3}{3}\left(\frac{1}{3}\right)^3\left(\frac{2}{3}\right)^0 = 1 \times \frac{1}{27} \times 1 = \frac{1}{27}$. These two possibilities are mutually exclusive, so the probability of choosing more yellow than red marbles in three draws is their sum, or $\frac{2}{9} + \frac{1}{27} \approx 0.26$. So it's likely that, if you were to perform this experiment many times, on close to 26% of the experiments, you'd draw the yellow marble more times. However, if instead there were six marbles in the bag, in the same proportions (four red and two yellow), the probability of choosing more yellow marbles than red marbles would be only about 10%. The probability of drawing yellow marbles at least four out of six times is the sum of the probabilities of choosing yellow marbles exactly four, or five, or six times:

$$\binom{6}{4}\left(\frac{1}{3}\right)^4\left(\frac{2}{3}\right)^2 + \binom{6}{5}\left(\frac{1}{3}\right)^5\left(\frac{2}{3}\right)^1 + \binom{6}{6}\left(\frac{1}{3}\right)^6\left(\frac{2}{3}\right)^0$$
$$= 15 \times \frac{1}{81} \times \frac{4}{9} + 6 \times \frac{1}{243} \times \frac{2}{3} + 1 \times \frac{1}{729} \times 1 = \frac{73}{729} \approx 0.10.$$

Thus when the marble "population" size is smaller, improbable outcomes such as drawing more yellow marbles more often are more probable. The same thing is true of biological populations. The fact that there are more red than yellow marbles in the bag is analogous to a fitness difference: red is "fitter" than yellow, but the chance of the improbable outcomes decreases as the overall number of marbles increases.

Wright-Fisher models can be modified in various ways. For example, when mating is nonrandom (§0.3.5), or population size fluctuates (see §2.1), the Wright-Fisher models described above are unlikely to provide good estimates. However, there are ways of calculating an "effective population size" to replace the population size parameter $2N$ in equation (0.3), so that inferences based on that equation (or related ones) provide better estimates. Different concepts of effective population size are needed for different contexts (Ewens 2004, 128), but the effective population size concept that's most common is known as "variance effective population size" (e.g., Gillespie 2004; Rice 2004).[25] Wright-Fisher models can also be modified in other ways, to model, for example, more than two alleles, or to incorporate mutation or migration (Ewens 2004).

Comparing figures 0.1 and 0.2, you might suspect, correctly, that if the population size for our Wright-Fisher model were increased further beyond 500, the curves would become even narrower than in figure 0.1. In the limit, as population size N goes to infinity, the probability distribution over frequencies in each generation would tend toward a distribution that assigns all probability to one particular frequency; in the limit there would be (almost) no possible variation in frequencies at any given generation.[26] However, I argue in a paper that is in preparation (Abrams InfPops MS) that it's incorrect to say, as some philosophers of biology do (Sober 2008; Morrison 2015; Potochnik 2017), that there are infinite-population models, that assuming an infinite population is an idealization, or that the idea of natural selection in infinite populations is part of the foundation of evolutionary biology. My view is that there are no models of evolution involving infinite populations, because evolution requires change in relative frequencies, and relative frequencies in an infinite set are either undefined or equal to zero. (No idealization can completely undermine all of its applications.) So-called infinite-population models are simply models that have no role for drift. Biologists do often *say* things about evolution in infinite populations, and these claims are usually correct: the role of this terminology in practice implies that it should not be understood literally.

0.3.5 OTHER INFLUENCES ON EVOLUTION

Evolutionary processes are complex. Although I focus primarily on natural selection in this book, and I briefly describe my view of drift in chapter 6, natural selection and drift are not the only dimensions of evolutionary processes that are important. Below I briefly mention other influences on evolution that will come up in various places in the book. They are also factors that my approach must ultimately accommodate, but I don't believe this will be difficult in most cases.

Linkage disequilibrium, *genetic linkage*, and *genetic hitchhiking* play enormous roles in contemporary evolutionary biology. I discuss them briefly in chapter 3, and I'll leave definitions for that chapter; see also Abrams (2015) and some unpublished work (Abrams InfPops MS). *Epistasis* occurs when the genes at different loci interact to produce an organism's phenotype, and the contributions of the two loci can't be decomposed into separate quantities of influence that can be added or multiplied to produce an effect of a given size. *Pleiotropy* occurs when one allele or genotype influences more than one phenotype. *Recombination* occurs when DNA subsequences from two different strands are swapped, for example, in paired chromosomes. (Section 2.1

mentions one kind of effect of recombination, chromosomal inversion, and recombination plays a role in linkage disequilibrium.) *Mutation* of course includes any processes by which DNA sequences are modified (other than by recombination; but cf. Abrams 2009b). We can treat *migration* of organisms between populations as influences on the evolution of those populations. More generally, the degree to which individuals in different parts of a population are more or less isolated from each other can affect the evolution of the population. *Mating preferences* can also result in individuals being more or less isolated from potential reproductive partners. *These* factors are part of the background of my discussion of a paper by Byars et al. (2010) in chapters 5 and 7 (also see Abrams 2009b, 2012a). A complete list of factors influencing evolution would probably be impossible to compile (§4.5).

Evolutionary biologists often measure relationships between parents and offspring in terms of *heritability*. There are, in fact, two main concepts of heritability in use, broad sense and narrow sense (Gillespie 2004), with variations and related concepts for particular contexts (Lynch and Walsh 1998, 2018). The two heritability concepts are mathematically different and are used in different ways, but both are defined in terms of correlations between parents' and offspring's traits. Generally speaking, these are measures of the extent to which parents' traits cause offspring to have similar traits. They are important, but I want to abstract from them to allow for the possibility of a broader range of inheritance relations between parents and offspring. I'm inspired in part by various ideas about *biased inheritance*, including arguments by Godfrey-Smith (2009b) and research on biasing effects in reproductive processes, such as segregation distorters (Charlesworth and Charlesworth 2010; McLaughlin and Malik 2017). Although not obviously a form of inheritance, *niche construction* (Odling-Smee, Laland, and Feldman 2003), in which organisms influence other organisms via alterations to the environment, or other *nongenetic influences* between parents and offspring or later descendants, may produce inheritance-like effects. I also won't discuss *cultural influences* in higher animals, except in the discussion of the Byars et al. study mentioned a moment ago. For the most part, when I talk about inheritance in this book, what I say can be translated into one or the other heritability concept, but I don't want to assume that everything relevant to inheritance can be captured by heritability relationships per se. Thus instead of talking about traits being "heritable" or speaking of "heritability," I'll often use the somewhat awkward substitutes *inheritable* and *inheritability*. I'll still use "heritable" and "heritability" when discussion of another author's work makes those terms appropriate, though. My loose description of heritability above will be sufficient for the points I'll make.

0.3.6 CAUSALISM AND STATISTICALISM

Among philosophers of biology (and a few biologists), there has been a debate in recent years about clusters of views that have come to be labeled "causalism" and "statisticalism." As mentioned in the preface, I'll sometimes frame my arguments in the context of this debate.

Causalists argue that natural selection, random drift, and other factors often labeled as "forces" (see above) are causes of evolutionary change (or, in some cases, of stasis). Viewing natural selection, for example, as itself a cause of change might mean viewing it as a general sort of higher-level cause that supervenes on details concerning populations: two different populations with different environments and different sorts of organisms could nevertheless be influenced by natural selection in the same way, because the direction, strength, and so on of natural selection doesn't depend on every detail of the population and environment that realizes it. On the other hand, causalists differ as to whether they take natural selection to be something that fundamentally applies to populations (e.g., Millstein 2006) or that can also be defined for pairs of competing organisms (e.g., Brandon 1990, chap. 1; Bouchard and Rosenberg 2004). Most causalists probably think that trait fitness derives from fitnesses of token organisms, and the propensity interpretation of fitness is the best known causalist view (§0.3.2), but it is not essential to the view. Causalists differ as to how to characterize the relationship between selection and random drift, but they have focused less on relationships to other evolutionary forces.

Statisticalists, by contrast, maintain that there are no general, higher-level causes of evolution that can go by the name of "natural selection," "drift," and so on (e.g., Walsh, Ariew, and Matthen 2017). Evolution takes place—frequencies of traits in a population change, for example—and this is caused, but the causes are merely events involving individual, token organisms and their interactions. On this view, there is no sense in which these lower-level causes, as such, aggregate into natural selection or drift understood as causal factors. For example, statisticalists sometimes claim that token organisms have fitnesses that involve causal properties of the organisms (e.g., Matthen and Ariew 2002; Walsh 2007), and in this respect, there seems to be agreement between statisticalists and advocates of the propensity interpretation of fitness. However, statisticalists deny that trait fitnesses can be defined in terms of these token-organism fitnesses in a way that would allow us to make sense of the idea of natural selection as a cause. Statisticalists do sometimes allow that the terms "natural selection" and "drift" can be used sensibly, but they argue that the terms should then be understood as mere summaries of effects—for example, certain changes of frequencies of traits in a population—rather than as causes.[27] In some cases, this implies that natural selection and drift can't be distinguished, or that they can be distinguished only when defined in terms of outcomes rather than kinds of influences. Statisticalists sometimes argue that explanations of evolution that abstract from details about individual organisms are based on mathematical models and that what is explanatory in these cases is purely mathematical.[28] I won't object to statisticalist arguments that certain models make it possible to explain some evolutionary outcomes purely mathematically, once the relevant mathematical assumptions have been justified by prior causal assumptions about a population (Ariew, Rice, and Rohwer 2015; Ariew, Rohwer, and Rice 2017). However, I don't believe this is the kind of explanation that is central to contemporary evolutionary biology. I develop a view of evolutionary processes as causes of evolutionary change that can support the view that causal explanations are central to evolutionary biology.

Some statisticalist arguments have roughly the following form (e.g., Matthen and Ariew 2002, 2009; Walsh 2007, 2010; Walsh, Ariew, and Matthen 2017):

Premise 1:

a. The choice between using model X or model Y is arbitrary (i.e., we can choose freely which to use); or
b. the choice between dividing up populations in way X or way Y is arbitrary; or
c. the choice between describing evolution as the result of selection or drift is arbitrary.

Premise 2:

Whether natural selection, drift, or both take place depends on the preceding arbitrary choice.

Premise 3:

Whether one thing causes another is a fact about the world that's independent of our choices; arbitrary decisions can't determine whether a cause exists or not.

Conclusion:

The putative causes (natural selection, drift, or both) are not causes.

That is, whether selection or drift takes place depends on an arbitrary decision, but causes can't vary depending on arbitrary decisions, so selection and drift are not causes. Causalists have usually responded by arguing that the relevant decisions are not, in fact, arbitrary (e.g., Gildenhuys 2009; Millstein 2009; Otsuka et al. 2011; Pence and Ramsey 2013; Strevens 2016). Chapter 6 explains why I think this is the wrong response.

0.4 Conclusion

In this chapter, I have, I hope, provided sufficient background on probability and philosophy of probability, and on evolutionary biology and philosophy of evolutionary biology, that the rest of the book will be intelligible to those without previous background in some of those areas.

PART I

Laying the Foundation

Population-Environment Systems

1.1 A Disagreement about Measurement

Imagine two distinct subpopulations of a larger population of, say, house sparrows, that evolved from a common ancestral population. Many house sparrows live in urban areas, so the subpopulations might consist, for example, of birds nesting in San Francisco and Oakland. Consider a particular genetic locus with two alleles. (For unfamiliar biological concepts please see §0.3 and its endnotes.) It's likely that the relative frequencies of these two alleles will differ in the two populations due to drift, natural selection, or other factors. It would be useful to be able to measure these differences between the two populations. Among other things, it could be useful to figure out whether the two populations' allele frequencies exhibited a pattern that would be probable given drift alone. Drift always occurs, in practice, so departures from what's likely given drift can be evidence that other evolutionary influences have been at work (see §2.1 for an illustration). One well-known way of quantifying this sort of difference between populations is with a statistic known as F_{ST}.

Building on work begun in the 1920s, Sewall Wright (1951, 1969) developed a set of mathematical concepts known as "F-statistics," or "fixation indexes," which measure various departures of gene frequencies from Hardy-Weinberg equilibrium. F-statistics such as F_{ST} can be used to measure nonrandom patterns in genotype frequencies resulting from a variety of factors. Depending on context and details of usage, F-statistics can be used to describe and provide evidence for various patterns of nonrandom mating, multilevel population subdivisions, degrees of isolation between populations, migration, and natural selection (Weir 1996; Weir and Hill 2002; Holsinger and Weir 2009; Weir and Goudet 2017; see also Ishida 2009). F-statistics consequently have close connections to ideas central to evolutionary theory, and they continue to be used in a broad range of biological applications, both theoretical and practical, across diverse species, using genetic data of various kinds (Weir and Hill 2002).

Inspired by Wright, and working during the same period, Gustave Malécot (e.g., Malécot 1969) developed closely related ideas about genetically related individuals, but Wright's and Malécot's assumptions were incompatible with each other (Ishida 2009), and there were some problems with Wright's claims about the measurement of F-statistics (Weir 2012). In part because of this situation, in the 1960s, 1970s, and 1980s, Masatoshi Nei on one hand (e.g., Nei 1973, 1977, 1987) and Clark Cockerham and Bruce Weir on the other (Cockerham 1969, 1973; Weir and Cockerham 1984) developed incompatible refinements of Wright's F_{ST}, with different ways of estimating it. Nei's variants of F_{ST} estimation measure genetic relationships between actual populations of related organisms. Cockerham and Weir's approach to estimating F_{ST} also involves measurements of actual populations, but their F_{ST} estimate is defined in relation to a large number of counterfactual "replicate populations" (Weir and Cockerham 1984; Weir 1996; Holsinger and Weir 2009).[1]

Nei criticized Cockerham and Weir's way of estimating F_{ST}. In a comment on work by Cockerham (1973), Nei and Chesser (1983) complained that Cockerham defined his estimator of F_{ST} "in reference to an imaginary population, as in the case of Wright's definition of fixation indices" (253). Nei also wrote that

in natural populations, however, each subpopulation usually has a unique evolutionary history, so that the concept of "replicate subpopulations" does not apply. (Nei 1987, 163–64)

Weir later wrote that

as a description of current frequencies, Nei's approach is appropriate, but there is then no basis for an evolutionary interpretation of estimates and no justification for making statements about divergence from ancestral populations, the effects of natural selection, or the extent of migration, for example. (Weir 2012, 639–40)

Thus, on one hand, Weir claims that Nei's approach to estimating F_{ST} is only suitable for describing differences between current subpopulations and should not be taken as evidence for evolutionary processes. On the other, Nei objects that Cockerham and Weir's approach depends on nonexistent subpopulations and does not reflect the "unique evolutionary history" of the subpopulations measured.[2] The idea that a mathematical measurement of evolving populations should take into account populations that don't exist, and never will exist, may seem, on the face of it, bizarre. Yet Cockerham and Weir's method has been, and remains, enormously popular in empirical research in evolutionary biology.[3] So what are these "replicate populations," and why should they matter to understanding the evolution of actual populations?

F_{ST} is usually defined in terms of distributions of alleles in populations. Generally, researchers don't have complete data about these distributions. Therefore they take random samples from actual populations. Weir calls this *statistical sampling* (Weir 1996; Holsinger and Weir 2009). One then computes a particular function of the distribution in the samples, where this function would approach the true value of F_{ST} as more samples were collected. (For additional details see Abrams [2012a].) The value of such a function is an estimate of F_{ST}. Similar ideas can be found in Nei's approach to F_{ST}.

In addition to statistical sampling, we can think of a population at a given a time as having sampled genes from a population at an earlier time (see §0.3.3 and §0.3.4). In practice, actual descendant populations usually won't have the same distribution of alleles at a locus as an ancestor population. This is due to randomness in the processes leading to later generations (drift) and other factors such as selection, which bias whether some alleles make it into subsequent generations. Weir and Cockerham call this *genetic sampling* (Weir and Cockerham 1984; Weir 1996; Holsinger and Weir 2009), extending the common notion of single-generation sampling of gametes or zygotes, and so on, to multiple generations. Note that members of the same species typically have common ancestors, and more generally, all living organisms do also. Therefore, for any populations present at a time—or for a population consisting of the entire species—there is some common ancestral population from which all have descended. Mutation of course generates some of the alleles in descendant populations as well. However, since mutation rates are usually very low, for many purposes we can ignore it.

Given this picture, we can see that there are many different populations that might have arisen from a single ancestral population. Each such population would contain a different collection of alleles that could have been sampled from the ancestral population. These are Weir's hypothetical replicate populations. One can think of them as resulting from replicating the experiment of allowing the ancestral population to evolve. Thus, data on members of the actual populations is the result of statistical sampling from populations that are already the result of genetic sampling. This is the basis of Cockerham and Weir's method for estimating F_{ST}, and it justifies the way in which it is mathematically different from Nei's method (Weir 1996; Abrams 2012a).[4]

How is it possible to calculate a mathematical quantity that takes into account an unknown sampling process from nonexistent, counterfactual populations? In Abrams (2012a), I explained that if one looks at the details of Cockerham and Weir's way of estimating F_{ST}, it turns out that what they assume is that, given an ancestral population, there is a probability distribution over possible population states. That is, starting from an earlier, ancestral population, there is a broad range of various possible ways it could evolve, with probabilities for each small range of population states. A current, actual population is a representative of this probabilistic process, in that it is one way—usually, a relatively probable way—that things could have turned out. The fact that we don't know the allele frequencies in the common ancestral population, nor what the probabilities of its subsequent evolution were, doesn't mean that those concepts can't be used to justify the way in which F_{ST} is estimated (Abrams 2012a). Probabilities of past population changes are in effect estimated from current data. This makes the mathematics of F_{ST} estimation different from that in Nei's method.[5]

What I want to focus on here are the probabilities that Cockerham and Weir assume to exist, and their relationship to hypothetical replicate populations. This chapter and the next will introduce two concepts that will help us to understand these ideas. I'll suggest that what I call a *population-environment system*, beginning from some initial state, has *causal probabilities* for various ways that it could evolve. Over the course of the book, I'll argue that, when biologists study evolutionary processes, what they are often doing is using an actual population as evidence for an underlying causal, probabilistic process that (in fact) gave rise to this population in its current state—but that nevertheless could have resulted in other population states. I claim that Weir's "replicate population" terminology merely provides a way of capturing this intuition with a concrete metaphor. Roughly speaking, Cockerham and Weir's F_{ST} can be understood as measuring differences between actual populations that represent different ways that the same population-environment system could have evolved.

My arguments in this book will help to clarify the basis for Cockerham and Weir's approach to estimating F_{ST}, and why those of Nei's criticisms highlighted above seem unwarranted. However, what I aim for is much broader than the elucidation of particular cases such as the disagreement between Weir and Nei. The view that I'll introduce is intended to allow us to develop a deeper understanding of what evolutionary processes are, in general, and how their character supports the way that research is usually done in evolutionary biology. I'll focus specifically on the role of natural selection in evolution, but the scope of the picture I introduce goes beyond that role.

The rest of this chapter introduces population-environment systems and some elements of this idea. In section 1.2, I argue that real-world evolutionary processes involve a high degree of what I call "lumpy complexity," and section 1.3 introduces the concept of a chance setup. Both of these are important to the idea of a population-environment system, which I introduce in section 1.4. In the next chapter, I'll argue that population-environment systems realize causal probabilities.

1.2 Lumpy Complexity in Real Evolutionary Processes

There is enormous complexity to the ways in which interactions between organisms and environments affect evolution. This idea will play several important roles in the book, so I want to start with a concrete description involving house sparrows to suggest the kind of complexity involved. No description as brief as the following could do justice to the complexity of real populations and environments. Nor should one expect to find similar descriptions in scientific papers or books, whose goals would not be served by such an accounting (cf. §2.1). Nevertheless, my description is based on a large body of research about house sparrows, and I consider it to be quite plausible. Many of the claims are based on Ted R. Anderson's (2006) work, but I'll cite additional works. I'll use some research on other bird species, too, following the common practice among biologists of making reasonable hypotheses about a species' properties from what's known about closely related species.

House sparrows (*Passer domesticus*) live on most continents in a wide variety of environments, both urban and rural. A house sparrow that finds food on the ground in an area exposed to the sky may give a call to alert other house sparrows, waiting until others arrive—sometimes even forgoing food if they don't. This is not thought to be an altruistic behavior; it allows birds to feed more efficiently, as they can share the task of visually scanning the sky for hawks and other raptors (Caraco 1981). Whether a particular instance of such food-calling behavior makes survival and reproduction more or less likely than alternative behaviors plausibly depends on various factors: How prevalent are raptors or other predators who might notice that the house sparrow is feeding? Did the call bring the calling sparrow to the attention of a predator? How much food is there to share with other house sparrows? How many house sparrows are there nearby? How easy is it for them to hear a call? Are other house sparrows upwind or downwind? Is there intervening foliage that will degrade the sound (Kirschel et al. 2009; Crozier 2010) or noise that will mask sounds (Hu and Cardoso 2009)? Are nearby house sparrows already satiated? Are they busy with other tasks, such as nest building, courting, or sitting on eggs? Would a male who is trying to attract mates to a desirable nest cavity have more offspring if he responded to the food call or remained to guard his nest cavity? Are there other areas with food that are safer, or do they have other dangers, such as nearby feral cats (Beckerman, Boots, and Gaston 2007)? Note that, although a sparrow may be able to eat more if other sparrows share scanning duties, there is also competition for food within a feeding flock, and aggressive encounters use up some potential feeding time. Some sparrows are more aggressive than others, so the benefit of sharing the meal with other sparrows depends partly on which sparrows respond to the call.

Here are some additional factors that affect the possible benefits of food-calling behavior. What kind of food is available in the exposed area? What nutrients are calling house sparrows likely to need for energy for activity, for feather production, for maintenance of feather coatings, for fighting para-sites, or for pigmentation? Male house sparrows can get substances from some foods that allow them to synthesize melanin used to create their black chest patches; these may influence attractiveness to females or play a role in aggressive interactions between males. However, melanin and related sub-stances can also be used for physiological processes that fight parasite infec-tions (Catoni, Peters, and Schaefer 2009; see also Olson and Owens 1998). Whether eating some substances or eating others is more or less beneficial for a given male house sparrow might depend on the season, the number and kinds of parasites that it has or will have, and the properties of other nearby house sparrows.

The preceding remarks ignored many aspects of house sparrow life, such as environmental interactions affecting development. For example, the in-tensity with which a house sparrow nestling begs, influencing parents' feed-ing behavior and subsequent nestling survival, seems to be the result of an earlier gene-environment interaction (Dor and Lotem 2009). Galván (2018) showed that, among Eurasian nuthatch nestlings, adult nuthatch alarm calls, normally an indication of predators, cause oxidative stress, an imbalance in certain chemicals. This in turn affects gene expression in a way that modifies melanin processing, resulting in less intensely colored feathers. It's possible that something analogous occurs with house sparrows.

Consider that many of the elements mentioned above are likely to exhibit large-scale spatial or temporal patterns. What foods are available and where they are located depend on plant growth patterns that vary across the land-scape, and that vary over time. Some plants grow best in some soils, at some heights, under certain weather conditions, and so on. Kinds of predators vary, and they depend on the presence of other prey, which in turn depend on what other plants and animals are present. Weather patterns can affect house spar-row survival; for example, Ringsby et al. (2002) showed that more nestlings die during bad weather on Norwegian islands. They also showed that the effects of weather depended on other environmental conditions, as the effects were different on nearby islands. Note that weather of course varies from week to week, from season to season, from year to year. Also note that house sparrows are known to live in—and perhaps move between—urban, agricultural, and other rural areas, and they live in areas that include a variety of human activities, any of which might make certain phenotypes, including behaviors, more valuable during certain periods of time (see Evans et al. 2009).

In general, most biological populations involve interactions between organisms all of which are different, even when they're members of the same species. However, as some of the remarks above illustrate, there are interactions with diverse members of diverse species as well. In addition, an environment includes abiotic entities—rocks, dirt, air, and so on—of different sorts in different configurations, which can affect the distribution of other organisms, or the ease of moving about in certain areas, and may include toxic or nutritious substances. (House sparrows sometimes eat pebbles, perhaps in order to get nutritious minerals.) Survival, mating, reproduction, and so on for each multicellular organism are influenced not only by interactions with things outside it but also by its physiological processes, which in turn depend on varying, complex entities—organs—composed of smaller entities—cells—whose functioning depends on interactions between their parts and surroundings (see, e.g., Mitchell 2009). Those interactions depend, ultimately, on molecular movements subject to the randomness of statistical mechanics (see, e.g., Abrams 2017). Plausibly, some of the variations that affect outcomes for an organism such as a house sparrow depend on cross-domain, multidimensional interactions between any of the sort of things mentioned above, including genetic, cellular, physiological, behavioral, and perceptual processes; these can include, for example, interactions between organisms involving food, aggression, disease, energy levels, reproduction, and landscape structure and weather. In Wimsatt's (1974, 2007) terminology, an evolving population exhibits an extremely high level of *interactional complexity*.

You might think that systems studied in, say, the statistical mechanics of gases are much more complicated than evolving populations, because they can incorporate interactions between millions of molecules, with different positions and velocities, and possibly with different masses and other properties that affect their interactions. By contrast, there are at most tens or possibly hundreds of thousands of organisms in a population. But the complexity of paradigmatic statistical mechanical systems pales by comparison to that of a biological population. In a gas, most molecules behave in almost the same way. The number of molecules involved in an evolving population in an environment is on the order of the numbers of molecules treated in statistical mechanics, but in the biological case these molecules are constrained to interact in ways that depend on entities at different levels—where the levels themselves have fuzzy boundaries. This makes evolving populations much more complex, at least as an object of study, than many paradigmatic complex systems in the physical sciences. I call the kind of complexity present in evolving biological populations *lumpy complexity*, because it involves clusters of properties— objects—that are somewhat similar but also dissimilar, interacting in complex ways across levels, and that often themselves are internally complex in a lumpy sense. (This is one reason it would be absurd to think that more than a few of the idealizations involved in models needed for empirical research in evolutionary biology could be removed or reduced; see §2.1.)

1.3 Chance Setups and Probabilities[6]

As I explained in section 0.2, the mathematical theory of probability gives us no guidance as to the source of probabilities relevant to events in the world. Those probabilities from which other probabilities can be calculated must come from outside mathematics—from contexts in which an interpretation of probability is applicable to a *chance setup* in roughly Hacking's sense:

> A *chance set-up* is a device or part of the world on which might be conducted one or more *trials*, experiments, or observations; each trial must have a unique *result* which is a member of a class of *possible results*. (Hacking 1965, 13; emphasis in original)

I'll refer to sets of possible results as *outcomes*.[7] Examples of chance setups include situations in which a coin is tossed, cases in which a person has a certain confidence in a conclusion, and quantum mechanical experiments. Certain properties of the chance setup make it the case that trials of the setup will realize properties that satisfy the axioms of probability. A statement of an interpretation of probability (§0.2.3) proposes or explains how members of a certain class of chance setups are able to do that. For example, according to some Bayesian interpretations of probability, it is mental states or brain states (Zynda 2000; Appiah 2017) or behavioral dispositions (Howson and Urbach 1993) that satisfy the axioms of probability. (Thus "chance" setups needn't involve chance in the sense of objective probability; see §0.2.2.) In some propensity interpretations, the properties of the chance setup allow it to realize dispositions of different strengths, and it is these dispositions that satisfy axioms of probability (Popper 1959; Giere 1973; Gillies 2000). Note that it's not the fact that we have developed or applied an interpretation of probability that creates the probabilities. Rather, probabilities exist in the world whenever a chance setup realizes properties that are sufficient for satisfaction of the axioms of probability.[8]

Although properties of a chance setup realize probabilities, only some of its properties do so. For example, in the case of objective probabilities for the outcomes of tosses of dice, the shape of each die and the densities of the material at points within it are plausibly among those properties that help to realize probabilities, as are various properties of a person who tosses the dice. Other properties of a chance setup can vary without changing the probabilities. For example, the fact that a metal die is made of one alloy or another per se makes no difference if the alloys are of equal density. Or, note that although the densities of material in a pair of dice help to realize probability, the various paths through the air that the dice take during different tosses might not count as determining probabilities for outcomes.

We can call those properties that realize probabilities—according to a particular interpretation of probability applied to a particular chance setup—the *probability-realizing properties* of the chance setup according to a given interpretation of probability. These are the properties of the chance setup in virtue of which it realizes probabilities at all, and those in virtue of which the probabilities have the particular numerical values they do. We can distinguish the latter as the *probability-quantity-realizing properties* of the chance setup.

1.4 Population-Environment Systems

As I mentioned earlier, an organism is usually a specific, concrete, physical object, and we have many such organisms, so it's quite natural to think of evolution as fundamentally tied to actual token organisms. Of course, many biologists and philosophers also see traits in themselves as playing a crucial role in evolution, and the central importance of environmental influences on evolution is widely acknowledged (§0.3). Nevertheless, I believe that a focus on populations of token organisms as such sometimes grips our imagination, and it constrains our thinking about evolution in ways that have been unhelpful. I see this focus as contributing to philosophical conundrums and seemingly irresolvable debates among philosophers and biologists.

In the following sections, I sketch a different way of thinking about evolutionary processes, viewing them as taking place in large, complex entities that I call "population-environment systems." An actual, evolving population of token organisms then embodies one path that such a system might have taken. One of my goals in this book is to show that thinking in terms of this conception makes it easier to fit together various facts about evolutionary biology. However, because I do think it's natural to view evolution as primarily about what happens to actual token organisms in their actual, immediate surroundings, and because I want to make a different way of thinking equally natural, I'll introduce the idea of a population-environment system using a series of analogies. The analogies are not, strictly speaking, essential to understanding what population-environment systems are, but I think that the complexity (§1.2) and abstract character of a population-environment system is more easily approached by comparison with familiar, or at least somewhat simpler, systems. After introducing the idea of a population-environment system in section 1.4.1, I discuss roles of traits in population-environment systems (§1.4.2) and the idea that population-environment systems are chance setups (§1.4.3).[9] Concepts reviewed in sections 0.2 and 0.3 are particularly relevant to this part of the chapter. Later, in section 2.3, I'll argue that population-environment systems realize causal probabilities.

1.4.1 ANALOGIES

0

I'll mention first that some aspects of the idea of a population-environment system were inspired by ideas from ergodic theory (e.g., Friedman 1970; Nadkarni 1998), a branch of mathematics originally intended to model physical systems in which a complex state of a system at one time is mapped, moment by moment, to subsequent complex states. A common illustration of this pattern is a cup of coffee into which cream is stirred; the state consists, for example, of points that are dark brown or white. The system evolves as the cream swirls into the coffee, with the state at one moment being mapped into the state at the next moment, until the entire liquid appears beige. Although very few of the basic ideas of ergodic theory are applicable to population-environment systems, it's worth noticing that all of the analogies below can be viewed as systems in which a state at one time is mapped into a state at a later time, which in turn is mapped to a state at a subsequent time, and so on. Some readers may notice other analogies with the coffee and cream example.

1

Consider a Galton board, or quincunx, a device invented by Francis Galton in 1873 (Kunert, Montag, and Pölmann 2001). A Galton board usually consists of a flat board on which pins are arranged in rows, with pins in each row

FIGURE 1.1 A Galton board, or quincunx, in the lobby of 19200 Von Karman Avenue, Irvine, California. Photo from https://galtonboard.com/gallery, with permission from Four Pines Publishing, Inc.

staggered relative to the previous one, so that pins in one row are halfway between those in the preceding row (figure 1.1). Balls are dropped into the pin array in the center (near the top of the figure), and they bounce from one pin to the next, finally dropping into slots in the bottom portion of the board. Galton boards are often used to illustrate a binomial probability distribution or a Gaussian distribution, because the way that balls pile up in different slots is thought to approximate such distributions.[10]

Two things are particular interesting about Galton boards from my point of view.[11] First, they show how the outcome for a given ball—the outcome of landing in one slot rather than another—is sensitively dependent on exactly how the ball is dropped. Even when balls are fed into the Galton board automatically from a hopper containing many balls, one still finds that the balls are distributed among the various slots at the bottom. The balls and pins can be as alike as we can make them, and the balls may be dropped in ways that seem indistinguishable, but because of minute variations in how the balls hit the pins, balls end up in different places. The second point of interest about Galton boards is that they typically produce a characteristic pattern of outcomes. Moreover, these patterns can be manipulated by altering properties of the pins, balls, and other characteristics of the device (Lue and Brenner 1993; Kozlov and Mitrofanova 2003; Arai et al. 2012). While it's not clear what interpretation of probability might apply to the individual ball slot outcomes, the fact that frequencies of outcomes (slots into which balls come to rest) follow common patterns, and that these patterns can be manipulated, suggests that one or more interpretation of probability should apply to these outcomes.

Note that when a single steel ball is dropped into the pin area and it subsequently bounces off many pins, there may be a probability as well as a frequency of the ball bouncing to the right conditional on hitting a pin. Such probabilities and frequencies could be altered, for example, by placing a strong magnet on one side of the Galton board so that balls are pulled toward that side. So within a *single* ball-dropping trial there are frequencies of bounces to the right and left (at different pins), which can be manipulated along with probabilities.

2

Now consider a pinball machine. This is a sloped box on legs, often with a colorful vertical display at one end. The box, known as a "playfield,"[12] usually has, at one end, a spring-loaded plunger that can be used to send a ball rapidly rolling up toward the top of the playfield. The ball then falls, bounces, or bumps into various posts, bars, holes, buttons, ramps, tubes, magnets,[13] and other devices, sometimes scoring points for the player as a result. When the ball comes back down toward the lower end of the box, it may come within range of flippers under the player's control, and the player will try to prevent the ball from "draining"—falling off the playfield into a ball reservoir—by hitting the ball with a flipper.

Pinball machines are like Galton boards in many respects.[14] Again we find balls bouncing off objects in complex ways that depend on minute variations in how the balls hit the objects. Although the movement of the ball is controlled to some extent by the player, what it does is still subject to sensitive dependence on earlier conditions—even for expert players.[15] For example, the difference between a ball that scores additional points and one that drops off the board without any new points can depend on exactly how a ball hits a bumper. Pinball machines can change over time as well. Playfields can wear down or become dirty over time, also affecting the balls' behaviors. In pinball machines with magnets, balls can gradually get magnetized, also leading to changes in behavior over time.[16]

3

There are some pinball machines that can be configured to play partially or completely on their own without the intervention of human players.[17] So imagine the following hypothetical super-duper pinball machine, which we might call an automatic Pinball Evolution (PE) machine. Balls are automatically released into an enormous, warehouse-sized pinball machine in response to events that occur on the board. Perhaps, for example, whenever a ball hits a particular bar, a new ball is released, with the new ball's initial velocity a function of the way in which the other ball hit the bar. Suppose further that there are balls of different sizes or materials, resulting in the balls bouncing in different ways, even when they hit the same object in the same way. Or perhaps different balls are influenced to different degrees by magnets under the playfield. Because balls can get magnetized over time, there may be developmental changes over the course of a ball's "lifetime." We can even imagine that some balls include gyroscopes, moving internal magnets, and various internal processes that affect how they interact with playfield elements at different times.[18] Assume that when a ball hits one of the special "reproduction" bars, a new ball of the same type is released—it is "born." When a ball drops off the lower edge of the playfield, it "dies."

Because playing fields can wear down, there may be changes in the balls' "environment" over time. Bumpers and posts also could be set up so that their properties change relatively quickly, even to the extent that some such devices disappear from the playfield—just as some bumpers and posts can pop up or down or turn on and off in response to events in real pinball machines. Other, new objects might arise on the playfield. Some of these changes might happen in response to interactions with balls. There need be no limit to the complexity of these automatic pinball machines.

Over time, balls of one kind may proliferate, while others gradually disappear from the board, giving us a pinball analogue of natural selection. Different pinball machines—the balls' environments—could be configured in such a way that they end up favoring the generation of more and more copies of one kind of ball, and fewer of another kind. We can also imagine that while new balls are usually copies of those that cause them to appear, occasionally a mistake is made, and a ball with a different size, elasticity, ability to be influenced by magnets, or internal mechanisms is produced. If this ball manages to hit a reproduction bar, a similar ball will be produced. Thus there is also inheritance of "mutations."

The automatic PE machine is obviously supposed to be an analogy for an evolving population in an environment, with complex interactions between organisms (balls) and their environment (the rest of the pinball machine). The idea of a PE machine emphasizes the ways in which the balls are simply part of a complex causal system, whose characteristics make some patterns of long-term change more likely than others—even though those changes are still under the "control" of whatever kind of chances result from effects of minor differences in earlier activity (see chapters 8, 9). The automatic PE machine is also, like the cup of coffee, a system in which states at one time are iteratively mapped into states at subsequent times.

4

Rather than thinking of evolution as something that happens to token organisms, or lineages, or even populations of actual individual, token organisms, I suggest that we think of evolution as what takes place in—or rather to—a large, very complex automatic machine—a "population-environment system."[19] Such population-environment systems are "constructed" by nature, of course, and they can involve an enormous range of complex and subtle physical, chemical, and biological interactions between both abiotic and biotic elements. The house sparrow illustration in section 1.2 only begins to suggest how much more complex a real-world population-environment system is in comparison to a super-duper PE pinball machine. Yet, here too, it is the properties of the whole population-environment system that determine its characteristic yet fluctuating patterns of outcomes.

5

One final analogy. The earth's weather system is also enormously complex. It incorporates elements that appear repeatedly, in some form, though different in each instance. For example, there are numerous types of clouds, which have characteristic heights and behaviors and form and precede each other in characteristic circumstances.[20] We can say something similar about organisms and the various environmental conditions that they encounter in their lives.

1.4.2 TRAIT REALIZATIONS AND TOKEN ORGANISMS

We should think of traits—organism types—in population-environment systems as properties that can be realized by the population-environment system. Of course, such a realization only happens when there is an actual organism with the trait within a population-environment system. We should think of population-environment systems as also having to do with certain *possible* realizations of organism types. For example, each token house sparrow within a single population-environment system is a realization of the house sparrow type in the population-environment system. Narrower types—such as particular genotypes, or having a large dark breast patch, or being able to process melanin in foods in certain efficient ways—are also types that may be realized, in particular locations and times, within a population-environment system as a whole. Each realization of an organism type occurs in conjunction with other characteristics—other genotypes, feather characteristics, stomach structures, and so on—and in relation to different conditions outside of the sparrow. Nevertheless, we can view the population-environment system as repeatedly realizing types, such as the house-sparrow-with-large-black-breast-patch type. Moreover, some of these realizations influence the system to produce similar realizations: as genes or other characteristics are passed from parents to offspring, the same or similar types are realized again, in part because of their earlier realizations. We can think of these influences as occurring between aspects of the population-environment system as a whole.

1.4.3 POPULATION-ENVIRONMENT SYSTEMS AS CHANCE SETUPS

Population-environment systems, I suggest (chapters 8, 9), are like Galton boards or run-of-the-mill pinball machines in that *particular* outcomes for particular parts of the system—balls or organisms—exhibit sensitive dependence on earlier conditions. This is consistent with the view that evolving populations are enormously complex. Population-environment systems are also like Galton boards, or my hypothetical Pinball Evolution machines, in that different systems can have different characteristic patterns of outcomes. Given what I've said so far about population-environment systems, I think it's reasonable to view a population-environment system as a chance setup in the sense defined above. (Wright-Fisher models, discussed in section 0.3.3, capture some central features of population-environment systems, despite the extreme simplicity of the models: there are probabilities of evolving in various ways, which depend on general characteristics of the population-environment system, some of which might be represented in a Wright-Fisher model.)

That population-environment systems are chance setups should be considered central to the concept of a population-environment system, with outcomes typically depending on enormously complex interactions—as in the house sparrow illustration above (§1.2). A trial begins with a state of the entire population and its environment at some chosen time (more on this below), and it continues until a specified later time. (For organisms with nonoverlapping, annual generations, a natural starting point could be a date and time in the spring, for example.) The outcomes are realizations, perhaps many times, of properties by organisms and possibly other aspects of the system. So rather than seeing outcomes—such as having a certain number of offspring—as something that happens to a particular token organism, we should view such an outcome as occurring for one or more realizations of a particular trait—realized, for example by the system as a whole via a component or aspect of the system. Of course, this component or aspect is precisely what we would normally identify as an organism.

As a chance setup, a population-environment system is defined in part by a set of probability-realizing properties (§1.3). These properties are implicitly and vaguely defined by biologists' decisions about what to study (chapters 5, 6). Some of the probability-realizing properties are specified by an implicitly assumed set of constraints on states of the environment and members of the population (chapter 6). For example, the environmental properties that might define a population-environment system could include a range of temperature variation, a specific topography, and the inclusion of certain species of animals and plants. The possible members of the population that define the population-environment system would usually be constrained to be members of a particular species or subspecies. Not every property of a population-environment is included among those that define its type as a chance setup, though. Chapters 4 through 9 discuss various dimensions of the fixed constraints that are part of the probability-realizing properties of a population-environment system.

The other component of the probability-realizing properties of a population-environment system is the initial state of *some* but not all facts about actual token organisms and their collective properties, including, at least, frequencies of some traits. (Two population-environment systems whose populations each included the same two competing traits *A* and *B*, although in different frequencies, and were identical in all other probability-realizing properties, would count as population-environment systems of different kinds.) However, actual events including states of actual organisms at *later* times are *not* usually part of what constitute a population-environment system. What actually occurs starting from the initial state of the population-environment system is just one possible realization of the system, consistent with the probabilities that the population-environment system helps to constitute. Thus, according to this default conception of a population-environment system, actual token organisms and actual frequencies do matter to what makes a particular population-environment system what it is, but *only* in a population-environment system's initial state. For other ways of constraining population-environment systems, see section 6.4 and various remarks in chapters 5 through 7.

Some readers may find it useful to think of a population-environment system as a process that is defined by a large, high-dimensional state space, with probabilities of the system developing along paths through different regions of the space.[21] What's included in the space depends on the constraints (implicitly) imposed by researchers' choices and on probabilities of various possible events. Things that can't happen starting from a population-environment system's initial state, or things that are inconsistent with other constraints imposed by its (implicit) definition, needn't be considered part of the state space. In addition, because the idea of a population-environment system is supposed to help us understand the foundation of evolutionary biology as practiced, events that have such incredibly small probabilities as to be irrelevant to evolutionary biology needn't be considered part of the state space. For example, the sudden appearance of a collection of one-hundred-foot tall, twenty-one-legged flying monsters due to an ultraimprobable confluence of quantum mechanical events is possible, but it's not something that is or should be relevant to evolutionary biology, so the state space for a population-environment system needn't allow this possibility. Nor should a population-environment system be thought to include the possibility that a population of bears might develop genes for neon blue fur, if the mutations that could lead to that event are so utterly improbable that they are not worthy of study for an earthbound evolutionary biology. Similar points apply to the possibility that strange weather patterns could lead to room-sized regions on the surface of the earth suddenly experiencing an absence of air molecules for hours at a time. On the other hand, if a philosopher or scientist found it useful to think of the state spaces of population-environment systems as incorporating such ultra-hyper-low probability events, that's OK, as long as we understand that these extended regions of the state space are irrelevant to evolutionary biology. (What matters here is not what evolutionary biologists say is relevant, but what is in fact relevant given what they are trying to do.)

At this stage, I think it may help some readers to think of population-environment systems as large, complex chance setups realizing (single-case) propensities for ways that a population might evolve. The idea that a population-environment system could realize propensities should seem reasonable to anyone who already thinks that token organisms realize propensities. After all, a population-environment system includes a collection of token organisms and token environmental conditions at an initial time t_0. So one could think of the chances of various ways that a population-environment system might develop as being analogous to, and supervenient on, the propensities for various organisms and elements of an environment to do various things under various future circumstances. Note however, that though this propensity view of population-environment systems can serve as a useful heuristic, I believe it is almost entirely wrong—as I argue in chapter 8. See chapter 3 as well.

1.5 Conclusion

I began in section 1.1 by summarizing a disagreement about the estimation of genetic differences between populations using F_{ST}, suggesting that ideas developed in this chapter and the rest of the book would help to clarify what is at issue. In sections 1.2, 1.3, and 1.4, I presented two core ideas that will play central roles in the rest of the book: the lumpy complexity of populations and environments, and population-environment systems as chance setups. After introducing causal probability and some other core ideas in chapter 2, I'll briefly come back to F_{ST} in section 2.3.

Causal Probability and Empirical Practice

This chapter introduces the concept of causal probability, which plays an important role in the book, and discusses modeling and statistical inference, along with arguments based on their use. I begin in section 2.1 with a brief introduction to roles of modeling and statistical inference in empirical research before introducing the concept of causal probability in section 2.2. In section 2.3, I give an argument that population-environment systems involve causal probability. The argument is what I call an *argument from empirical practice*, and section 2.4 discusses the general form of such arguments.

2.1 Models, Statistical Inference, and Reality

I'll use a recent example of empirical research to introduce some common roles for models and statistical inference in evolutionary biology. Among other things, the example will illustrate the role of frequentist statistical methods, which are pervasive in research on microevolutionary processes. I'll give further remarks on frequentist statistics in section 2.4, but I won't go into detail on long-running discussions about the value of different statistical methods (including frequentism, Bayesianism, likelihoodism, and other variants),[1] nor on the substantial literature in philosophy of science about scientific models.[2]

Santos et al. (2016a) studied the evolution of chromosomal inversions in the fruit fly *Drosophila subobscura*. (A chromosomal inversion occurs when a segment of DNA within a chromosome is reversed in some organisms; afterward, DNA sequences within the segment that had been nearer to one end of the chromosome are closer to the other.) Santos et al. wanted to know whether there was natural selection for or against some inversions. There are various reasons this kind of selection might occur (Dobzhansky 1970; Hoffmann and Rieseberg 2008; Kirkpatrick 2010; Santos et al. 2016a). The Santos team tracked changes in frequencies of chromosomal arrangements over a couple dozen generations in three laboratory populations of flies derived from a natural population in Portugal. The researchers used a somewhat complex Wright-Fisher model without fitness differences (see §0.3.4 for a simple version) to evaluate whether the observed patterns of changes in frequencies of chromosomal arrangements would be likely without natural selection. The idea is that a Wright-Fisher model without selection shows what sorts of patterns would be probable with drift alone. If the model showed that some of the observed frequencies were improbable given drift alone, that would count as evidence that those patterns of chromosomal arrangements—the frequencies in the populations—were partly due to natural selection. I won't describe Santos et al.'s conclusions in detail, but I note that the researchers did find evidence for natural selection on some inversions.[3]

Based on what Santos et al. (2016a) knew about laboratory populations of fruit flies, their Wright-Fisher model needed to represent effective population sizes (§0.3.4) as changing in each generation. Their model also differed from the simple models in section 0.3.4 because rather than modeling two alleles at a locus, Santos et al. needed to model multiple alternative DNA sequence arrangements. These differences from simple Wright-Fisher models made purely mathematical calculations of probabilities of chromosomal arrangement frequencies too difficult. As a result, Santos et al. used computer simulations to estimate the probabilities of DNA arrangements in different generations.

For each of the three laboratory populations, the authors completed five thousand simulation runs, all beginning from an initial state that reflected frequencies of DNA arrangements at the fourth generation in a particular laboratory population (i.e., one composed of real flies). To create each subsequent generation in a simulated population, representations of DNA arrangements were randomly sampled according to probabilities derived from the frequencies of arrangements in the previous generation (see equation (0.5) in §0.3.4). Typically, each of the five thousand simulation runs for a given laboratory population would produce a different pattern of frequencies of simulated chromosomal arrangements at the end of the simulation (the twenty-eighth generation). Each set of five thousand outcomes could then be used as an estimate of what a mathematical Wright-Fisher model would specify as probabilities of final frequencies of chromosomal arrangements.[4] (There are two kinds of frequencies here. The first is the frequency of a given chromosomal inversion at the end of one simulation run. The second is the frequencies, or rather distribution, of inversion frequencies across all five thousand simulated populations. Compare: if you want to know the chance of getting four heads in a row when tossing coins, you could multiply $1/2 \times 1/2 \times 1/2 \times 1/2 = 1/16$—which is easy, in this case—or you could toss four coins a few thousand times and then note that four-head sequences appear about 1/16 of the time.) Santos et al. (2016a) then compared the five-thousand-run probability distribution for each inversion with the frequency of that inversion in the corresponding laboratory population. If the frequency in the laboratory population was very improbable, according to simulation results that reflected drift alone, that was treated as evidence the inversion had probably been under selection. (I discuss this point below and in chapter 7.)

This example illustrates the idea that models can play an important role in statistical inference concerning evolutionary processes in real populations. Models provide simplified, approximate representations of possible probabilistic and causal relationships. Statistical methods can then be used to decide whether the model represents the actual properties of the population with sufficient accuracy. Several points are worth emphasizing.

First, note that the point of the modeling was to estimate what would be probable without selection on inversions. (In statistical terminology, the point was to estimate p-values for the data from the laboratory populations [see §2.4.2, §7.2.3]. A p-value in this case is the probability that the data would have been observed given the *null hypothesis* that the populations had been subject only to drift. That natural selection took place was treated as the *alternative hypothesis*.[5]) Santos et al. assumed that the model would inform them of what would happen in a possible population that would be similar to the actual population (see Huneman [2012, 2013, 2015] for more discussion of counterfactuals about evolution).

Second, to inform researchers about such a possibly counterfactual population seems to require that models *represent* such populations. Scientists do refer to models as representing, but the nature and existence of relevant representational relations and ways that they do or don't depend on some kind of abstract similarity are topics of active discussion by philosophers of science (Weisberg 2013; Gelfert 2016). In this book, I'll merely assume that it makes sense to talk about models representing actual or possible populations, and that some sort of abstract similarity between models and possible situations plays a role in how we use models.

Third, this example illustrates the fact that in some cases there is continuity between mathematical modeling and modeling using computer simulations. This is shown by the fact that Santos et al. (2016a) used a Wright-Fisher simulation rather than a Wright-Fisher mathematical model simply because the former was more tractable. More specifically, often one can use either mathematical models or computer simulations to derive similar conclusions. When computer simulations are the only option, the simulation model may be very similar to other simulation models that could have been replaced by mathematical models.

Fourth, there is much about Santos et al.'s (2016a) laboratory populations that the Wright-Fisher model necessarily ignores. Of course, the researchers controlled quite a bit in the laboratory environment. For example, flies that were designated as members of the same population were kept in several vials, with about fifty adult flies and eighty eggs per vial.[6] Then, about every twenty-eight days, the researchers would mix the six hundred to twelve hundred flies that were considered to be members of a single population. Despite this regimentation and other systematic procedures, there still could have been variation within and between vials—for example, in male flies' courtship behavior and female responses to it (Steele 1986a, 1986b). The model didn't track these details. Similarly, not all details of recombination, mutation, and effects of phenotypes on reproduction were studied (see Santos et al. 2016b). Ignoring many details of real-world processes is normal in evolutionary biology, as well as many other areas of science.

Fifth, a model can take into account some of the ignored factors by, for example, averaging over their likely effects. Santos et al. (2016a) adjusted effective population sizes used in their simulations based on genetic tests of the study populations and similar populations in earlier studies (see also Berthier et al. 2002; Simões, Pascual, et al. 2008; Santos et al. 2012). However, nonrandom mating is one factor that is relevant to effective population size. So, by adjusting the effective population size in the simulations for each generation's samples, Santos et al. may have adjusted the model in a way that would compensate for nonrandom mating in their experimental populations.

Sixth, we needn't assume that by adjusting the model to take into account ignored factors, the model thereby perfectly summarizes facts about the populations. Even as a summary of unspecified factors, the model is likely to be only be approximately correct. Further, researchers may sometimes incorrectly assume that unmodeled factors can be ignored. This could result in a model that doesn't fit what happens in the population very well, or that fits, but for the wrong reasons. When the model fits for the wrong reason, later research might show this to be the case (but see Wimsatt [1980b, 2007] on biases in models that can hide problems).

A general idea (from Wimsatt [2007] and Potochnik [2017]) that summarizes many of the preceding points is that often the purpose of models and statistical inference is to provide imperfect evidence for imperfect causal patterns. That is, models and statistical inference *provide imperfect causal patterns' imperfect evidence* (PICPIE):

PICPIE

Models and statistical inference provide

1. evidence that is imperfect (because it comes from methods that don't guarantee success and because models include idealizations[7]),

which is

2. evidence for the existence or absence of *causal patterns* (Potochnik 2017) in the world that are of interest to researchers, within the systems studied.

These ideas seem to describe most empirical research in evolutionary biology, as well as a great deal of other scientific research. I'll discuss each of the two components of PICPIE in turn.

The first component of PICPIE is the claim that, often, the degree of evidence is due both to the fact that statistical inference doesn't guarantee its conclusions and to the fact that the models involved only accurately represent, to a certain degree, some aspects of the systems modeled. The systems represented may be real or may be counterfactual systems referenced in the process of statistical inference (discussed below and in chapter 7). Note also that the degree of accuracy can depend on the goals and constraints of particular research endeavors (Potochnik 2017).[8] (In the case of Santos et al. [2016a], the researchers were interested in evidence for or against hypotheses that specific chromosomal arrangements had been under selection, so their inferences were designed to be somewhat accurate about those hypotheses, but not about all other aspects of the populations studied.) Various other factors can affect the accuracy of inferences; for example, research funding that affects the sort of data one is able to collect (Abrams 2012b), the fact that certain statistical methods have been developed,[9] or the availability of particular instruments (Bickle 2018; Craver 2022).[10]

The second component of PICPIE says that causal patterns are what the evidence is for. I use Potochnik's (2017) concept of causal patterns: these are regularities concerning what caused what, realized locally in some parts of the world, that need not be exceptionless. What counts as an instance of a particular causal pattern can be a matter of degree, and which causal patterns certain models and research are supposed to approximate depends on researchers' interests.[11] In the Santos et al. (2016a) case, the causal patterns would be the presence or absence of natural selection causing some of the frequencies of inversions to be larger or smaller. For example, the Wright-Fisher models Santos et al. used misrepresented many things about the populations they studied, but the experiments were designed primarily to gain evidence about particular causal patterns of interest—that is, about whether natural selection on particular inversions had occurred. Note that researchers' interests can determine both which causal patterns they want to learn about and the strength of the evidence they are trying to gain about the causal patterns.

Other parts of this book will help to motivate the view that empirical research of the kind described by PICPIE is common in evolutionary biology, and they will fill in details about the nature of the causal patterns studied. Note, among other things, the PICPIE pattern of research is required by the lumpy complexity (§1.2) of evolving populations. Because of this lumpy complexity, there is no reason to expect that biologists will be able to remove most of the imperfection in the evidence in the end, or that precisely satisfied causal patterns can be identified.[12] (Research in evolutionary biology is not less valuable for that.)

Another common role for models in evolutionary biology is to investigate plausible but hypothetical causal patterns in evolution. This involves "hypothetical pattern idealization" (Rohwer and Rice 2013), in which a theoretical model (such as a Wright-Fisher model) is used simply to illustrate the possibility of a kind of causal pattern that might play a role in a domain (e.g., Nagylaki 1992; Ewens 2004). Such models can figure in what are known as "how-possibly explanations" (Brandon 1990; Brainard 2020), "candidate explanations" (Huneman 2014), "possible explanations" (Grim et al. 2013), or "why-maybe explanations" (Richerson and Boyd 2005). These are explanations in which a possible causal process that could produce an outcome is identified, without specific evidence that the causal antecedents required by the model are realized. However, as Santos et al.'s (2016a) use of a Wright-Fisher model illustrates, models can also end up playing a crucial role in justifying more specific conclusions.

2.2 Causal Probability

This section introduces a concept of *causal probability* that I will use throughout the book. For review of basic philosophy of probability concepts, see section 0.2. For the concept of a "chance setup" and related specialized terminology below, see section 1.3.

2.2.1 CAUSAL PROBABILITIES[13]

A *causal probability distribution* is a set of probabilities realized by properties of a chance setup, where alterations of probability-quantity-realizing properties

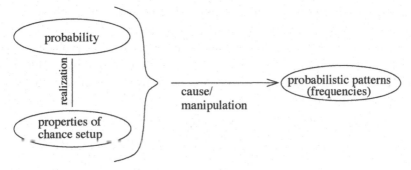

FIGURE 2.1 Causal probability relations. Originally published in Abrams (2015).

would usually change frequencies of outcomes (in numerous trials) in a way that accords with the change in the probabilities: those outcomes that would be more probable occur more often, most of the time, and so on (figure 2.1). For example, the objective probabilities that we attribute to tossing of dice appear to be causal probabilities since, roughly, changing densities within the dice changes probabilities and frequencies of outcomes in similar ways. Thus "causal" in "causal probability" refers to a difference-making (Lewis 1973) or manipulability (Woodward 2003) sense of "cause." My language about relationships between causal probabilities and frequencies is quite vague. That's because there are deep philosophical problems about these relationships, but as far as I know, no one has yet given a satisfactory solution to these problems (see, e.g., Berkovitz 2015). The vague relationships are nevertheless taken for granted in scientific practice. Note that the fact that we use frequencies as evidence in our attributions of probabilities to the tossing of dice doesn't change the fact that the probabilities, if they are objective, usually bear the right relationships to frequencies.[14]

One way that altering probability-quantity-realizing properties affects frequencies occurs when there are single-case causal probabilities with independent trials. For example, on the assumption that tosses of dice are independent, we can derive probabilities of frequencies in sets of many tosses, and it is these derived probabilities that are supposed to be close to frequencies of outcomes. However, the concept of causal probability doesn't require that the relationship between the probability-quantity-realizing properties and frequencies be derived from probabilities on multiple, independent trials. The relationship can be more direct, as in long-run propensity interpretations (§0.2.3). Moreover, we can also talk of causal probabilities where a single trial generates conditional probabilities for many events, with frequencies among those events. I illustrated this point in section 1.4.1 in my discussion of a Galton board. I see population-environment systems in the same way.

Note that the term "causal probability" and related terms such as "causally probabilistic" don't refer to any particular interpretation of probability. Probabilities defined by some (but not all) interpretations of probability are always causal probabilities. I call such interpretations "causal probability interpretations." All causal probability interpretations are objective probability interpretations, but not all objective interpretations define causal probabilities. The concept of causal probability therefore allows us to classify interpretations of probability into those that always define causal probabilities and those that don't. For example, as I argued in Abrams (2015), it's plausible that some propensity interpretations are causal probability interpretations, while Bayesian interpretations are not. Simple finite frequencies aren't causal probabilities, because changes in properties realizing probabilities don't cause changes in frequencies; they simply *are* changes in probabilities. Best system interpretations are also, prima facie, not causal probability interpretations, since they are defined directly in terms of frequencies (but there may be room for disagreement on this point).

In section 2.3, I'll argue that the way in which certain simulation models are used in evolutionary biology provides a good reason to think that evolutionary processes depend on causal probability of some kind. In section 7.2, I will argue that the role of statistical inference in empirical studies provides an additional reason for this conclusion. I'll discuss possible sources for causal probability realized by population-environment systems in chapters 8 and 9.

2.2.2 HUMPHREYS' PARADOX

Humphreys (1985) raised a problem for propensity interpretations (see also Salmon 1979; Lyon 2014), but a similar point applies to any causal probability interpretation: if chances are supposed to capture a causal relationship, expressing that relationship as a conditional probability of an effect E given its cause C seems to imply that effects can cause their causes. The problem can be illustrated using simple forms of Bayes' theorem (e.g., Grimmett and Stirzaker 2001):

$$P(C|E) = \frac{P(E|C)P(C)}{P(E)}.$$

Starting from the right-hand side, we see that from the probability $P(E|C)$ of an effect E conditional on a cause C, along with the probability $P(C)$ of the cause and the probability $P(E)$ of the effect (which incorporates all possible causes of it), one can derive the probability $P(C|E)$ of a cause given an effect. Suppose that all of the probabilities on the right side above are causal probabilities. Then the probability on the left would seem to be a causal probability as well. But that implies that we can manipulate frequencies of causes by altering probabilities of effects!

One way to avoid this problem is define objective probabilities in ways that don't allow treating certain cause-effect relationships as normal conditional probabilities. If probabilities of outcomes on a trial are not conditional probabilities of effects given properties of the chance setup, Bayes' theorem can't be used to show that properties of the setup have probabilities conditional on outcomes.[15] We can still apply Bayes' theorem to other conditional probabilities, though.

For example, suppose we have a biased coin C and two biased coins E_1 and E_2 that we toss—or, better yet, that we spin on their edges by flicking an edge with a finger, waiting to see which side is up when the coin falls. (It's difficult to bias a tossed coin, but not difficult to bias a spun coin [Gelman and Nolan 2002]). Here is our chance setup: We spin, or "flick," coin C, which gives heads a probability of 0.6. If the coin comes up heads, we flick coin E_1, which gives heads a 0.7 probability. If C is tails, we flick coin E_2, a fair coin whose probability of heads is 0.5. The outcome of the entire two-part trial is the *pair* of outcomes from the two flicked coins, but we can still consider the set of outcomes such that, say, the outcome of the first flick is heads and the outcome of the second flick is either heads or tails. We can also consider conditional probabilities of an outcome on one flick given an outcome on the other. From the probabilities specified, it's possible to calculate that the probability of heads on the first flick given heads on the second flick is equal to about 0.68.[16]

We can recognize this as a conditional probability of a cause (flicking the first coin) given an effect (flicking the second coin). However, this conditional probability is realized by properties of the entire two-flick setup, as are all of the probabilities used to calculate it. We can manipulate these probabilities, and frequencies in repeated trials, by manipulating the setup—say by changing the biases of the coins. As a probability defined by this setup, the conditional probability of heads on the first flick given heads on the second flick is unproblematic, but it is a causal probability relative to the entire two-flick setup, and there is no probability *conditional* on the setup per se. (As I note in section 2.3, my definition of population-environment systems in section 1.4 supports this this strategy.)

2.3 Why Think It's Causal Probability?[17]

I claim that evolution in population-environment systems depends on causal probabilities, because population-environment systems are chance setups that realize causal probabilities. The main justification for this claim is that,

I argue, a large body of successful empirical practice in evolutionary biology *implicitly* takes for granted and depends on the existence of what I call "causal probabilities." In this section, I'll use Santos et al.'s (2016a) laboratory study (§2.1) as an illustration of one common pattern of research involving simulations. This pattern seems to require causal probability for its success. I gave a similar argument in Abrams (2015), using an example of simulations in empirical research on human evolution over recent millennia (Voight et al. 2006). After presenting an argument for causal probabilities in this section, I discuss, in the next section, the general structure of arguments from empirical practice such as the one I give next. Later, in section 7.2.2, I give a different argument for the existence of causal probabilities in population-environment systems.[18]

Recall that Santos et al. (2016a) used computer simulations to model evolution in laboratory populations of *D. subobscura* fruit flies. In particular, the researchers used the simulations to estimate probabilities of frequencies of chromosomal inversions in their laboratory populations, on the assumption that the inversions were not subject to natural selection. If simulations showed that the probability of an *observed* pattern of inversion frequencies was low on the assumption of no selection, this would be evidence that selection had helped to produce the pattern.

My argument that the research of Santos et al. depends on causal probability has the following form:

1. Each simulation run realizes causal probability, on the assumption that the pseudorandom number generator (PRNG) code used in the simulation realizes causal probability.
2. The way in which the simulations were used to model possible real-world populations means that the populations also realize causal probabilities, on the assumption that Santos et al.'s strategy is a generally effective one.
3. Because the use of this strategy is very widespread in a science that has apparently been giving us insights into the nature of evolutionary processes, the strategy is effective. (That concludes the argument based on the assumption that the PRNG code realizes causal probability.)
4. Finally, even if the PRNG code doesn't realize causal probabilities, the design and behavior of the PRNG code, and its role in the simulations, was such that the first two parts of the argument, suitably modified, nevertheless justify their conclusion.

As in section 2.1, I'll use "frequencies" in two senses. Each run of a simulation reports numbers that represent frequencies of inversions. Santos et al. (2016a) performed five thousand simulation runs for each of three laboratory populations, and each set of five thousand simulation runs provided frequencies of simulation outcomes. Those outcomes are the numbers representing frequencies of inversions. Thus, a set of five thousand simulation runs provides frequencies of numbers that represent inversion frequencies. It is the former kind of frequency—frequencies in the five thousand runs—that scientists used to estimate the probabilities of the frequencies of the second kind.

The first point is that Santos et al.'s (2016a) *simulations* themselves realize causal probability, or at least model it in a very direct way. Each simulation run was different because the simulation code used a PRNG to determine values of variables during the run. Let's assume, initially, that an instance of PRNG code realizes the same chance setup each time it produces a number. Then when other code in the simulation code asks the PRNG code for a random number, and the PRNG code returns a number, this would count as a trial of the chance setup realized by the code. The PRNG code is designed to give such numbers known probabilities (see below for qualifications). In Santos et al.'s simulations, such numbers were repeatedly generated and used to determine steps in the operation of a slightly complex Wright-Fisher simulation (Santos et al. 2016b).[19] The idea is that Santos et al. did not know, initially, what probabilities simulation outcomes (inversion frequencies) would have, given the simulation code and the role of the PRNG code in it. So, while invocations of the PRNG code are trials of a (relatively) simple chance setup, with known probabilities, when an entire simulation run is performed, a trial of a complex chance setup with unknown probabilities occurs. It is the outcome probabilities of this more complex chance setup—that is, of the simulation—that Santos et al. didn't know. They then used the relative frequencies of simulation outcomes across sets of five thousand runs to estimate the probabilities of those outcomes. However, these frequencies were manipulable by changing the parameters of the simulations, and in fact, Santos et al. (2016a) used different parameters for each of three sets of five thousand runs, each set being designed to model a single laboratory population. So, by manipulating properties of the simulations, Santos et al. manipulated frequencies of simulation outcomes, and they used the resulting frequency distribution to estimate probability distributions for outcomes produced by the simulations. On the assumption that Santos et al. were correct to do this, each simulation realizes causal probabilities for its outcomes.

The second point is that the purpose of using multiple runs to estimate probabilities in the simulations was actually to estimate probabilities *in the laboratory populations*. Or rather, it was to estimate what the probabilities in the lab populations would be if there were no natural selection on inversion frequencies. That is, causal probabilities in the simulations were used to model probabilities in (possible) real-world processes. Thus, an implicit assumption was that probabilities in the lab populations could, in principle, have been manipulated in the same way as probabilities in the simulations. That is, the lab populations might have been subject to drift alone, or they might have been subject to natural selection, as well, for a given set of competing inversion frequencies, and this difference would usually affect relative frequencies of outcomes in ways that would correspond to differences in probabilities.[20] However, since the laboratory populations are difficult to manipulate experimentally, while simulations are easy to manipulate, scientists can make inferences about probabilities in laboratory populations by estimating probabilities in simulation chance setups.

So Santos et al.'s (2016a) study implicitly depends on an assumption that, in the simulations, probabilities and frequencies are manipulable together—so they are causal probabilities—and these causal probabilities are used to model lab population probabilities. The fact that causal probabilities in the simulations are being used to model and learn about probabilities in the lab populations is reason to think that Santos et al. also implicitly viewed the probabilities in lab populations as having the properties that define causal probability. One way to see this is to realize that it's plausible that if it had been easy to run the experiment with five thousand lab populations, the simulations wouldn't be needed, since probabilities of outcomes in the lab populations could have been estimated from frequencies in experiments. However, in addition to the large cost and amount of time that an experiment with thousands of fly populations might require, for fruit flies it would be impossible in practice to duplicate genetic and other conditions from one population to the next.

The third point is that there are many studies that use simulations in the same way that Santos et al. did—namely, to estimate possible probabilities for actual populations using multiple simulation runs to estimate those probabilities.[21] That is, this is a successful, established way to learn about whether natural selection, for example, has occurred in real populations. So it's implausible that it would be successful if the world didn't more or less fit with the assumptions of the research. Such research thus provides some reason to think that evolutionary processes involve causal probabilities.

Fourth, the argument above depends on the assumption that PRNG code constitutes a chance setup. That assumption is the basis of the argument that the simulation code also constitutes a chance setup, with each simulation run being a trial of the setup. However, if the assumption that the PRNG code constitutes a chance setup is incorrect,[22] the simulations were nevertheless being used exactly as if the PRNG code had been replaced with a source of causal probability. PRNG algorithms are carefully designed to produce behavior that is very similar to the behavior of chance setups (e.g., Knuth 1998; L'Ecuyer and Simard 2007; Kneusel 2018). (In particular, most PRNG algorithms are explicitly designed to behave largely as if outcomes were drawn from a large set of rational numbers between 0 and 1 that have equal probabilities on independent trials.) Even if the PRNG code doesn't constitute a chance setup, it is a model of a chance setup, modeling probabilities in a hypothetical, real chance setup embedded in a hypothetical simulation. In effect, if the simulation is not a chance setup, it would be a model of a hypothetical simulation that is a chance setup. And via this hypothetical simulation, the actual simulation models a population-environment system. The argument in the preceding paragraphs then goes through, since the method used by the scientists would not have been successful if the world did not behave as if it had causal probabilities, just as the intermediate, hypothetical simulations modeled by the real simulations do.

Now that I've argued that population-environment systems realize causal probability, I want to address Humphreys' paradox. A population-environment system does not determine any chances of outcomes prior to its initial state, only chances of subsequent outcomes. Probabilities relative to the initial state alone are *unconditional* probabilities. Chances of earlier outcomes must be relative to a different population-environment system, one with an earlier starting time.[23] Without these restrictions on the concept of a population-environment system, we would have a version of Humphreys' paradox (§2.2.2). However, as with the multiple-coin example in section 2.2.2, there is no special problem with conditional probabilities involving two later states that might occur during the operation of a population-environment system. This is so even for the probability $P(S_c|S_e)$ of a cause state S_c conditional on its effect state S_e, as long as this is understood as an overall population-environment probability—that is, a probability relative to the initial state of the population-environment system and the system's general constraints. (This will be important in §6.4.) Where it's useful to have a notation for probabilities relative to different population-environment systems ρ, one can distinguish this relationship from conditional probability by using a subscript $P_\rho(S)$ (e.g., Berkovitz 2015) or a semicolon $P(S; \rho)$ (e.g., Mayo 2018) rather than using the standard conditional probability notations $P(S|R)$ or $P(S/R)$.

Coming back to section 1.1's discussion of the role of probabilities of "replicate populations" in Cockerham and Weir's method for estimating F_{ST}, we can understand their picture as one in which the assumed ancestral population is the start of a long-running population-environment system. The probabilities of alternative replicate populations that figure in the derivation of an F_{ST} estimate are causal probabilities realized by this population-environment system. Only one possible state of the population-environment system can be realized in the present, however, perhaps incorporating multiple subpopulations, as it turns out. The current state is, most likely, one that incorporates causally probable allele frequencies given past facts about the population-environment system—whether there was selection, genetic linkage, migration, or some other biasing process. However, it is the existence of the causal probabilities that justifies Cockerham and Weir's method of estimation.

2.4 Arguments from Empirical Practice

The preceding argument is what one might call an *argument from scientific practice*. Such arguments are common in philosophy of biology and other areas of philosophy of science (e.g., Bogen and Woodward 1988; Batterman 2002; Wilson 2006; Wimsatt 2007; Brigandt and Love 2012; Sterrett 2014; Potochnik 2017; Waters 2017, 2019). Some of my arguments from practice, such as those in chapters 3 and 4, depend on general facts about roles of fitness concepts and the concept of natural selection in evolutionary biology. One might also think of these as arguments from theory rather than practice, but the theory that interests me is embedded in the practice of biology. Arguments like the one in the previous section are more specifically what one can call *arguments from empirical practice* (chapters 5, 6, 7). I use arguments from empirical practice not only to make claims about the epistemology of science but also to make metaphysical claims about characteristics of the world that allow evolutionary biology to succeed. Though this is not a new idea (see, e.g., Waters 2014), using arguments from practice to establish metaphysical claims puts extra weight on them, so it's worth spelling out the general form of the reasoning involved, as I do in section 2.4.1. In section 2.4.2, I discuss the possibility that recent evidence of systematic biases and errors in some scientific practice might undermine my arguments. I also discuss a general concern that some readers may have about uses of p-values in empirical research.

2.4.1 STRUCTURE OF THE ARGUMENTS

Arguments from empirical practice can be summarized like this:

> The approximate truth of assumption A would provide the best explanation of the success of many scientific results in domain D, given the role of method M in those results and its dependence on assumption A.

For example, in the argument of the preceding section, A is "population-environment systems realize causal probability," M is the method of using computer simulations to estimate probabilities in the world, and D is evolutionary biology. Note that the "assumption" A doesn't have to be one that the scientists using method M explicitly acknowledge, and additional work may be needed to establish the dependence on A. Here is a more detailed outline of arguments from empirical practice:

1. Argument that research method M is probably approximately truth-conducive:
   a. In a particular scientific domain D, if a widely used research method M often misled scientists, and did not help scientists learn about the world, it's likely that they would realize that the method was producing unhelpful results. That is, the fact that a method is in wide use over a long time is evidence that the claims inferred using this method are approximately true.
   b. Many scientists have published results in domain D using method M, many times, over many years. Typically, this means that there has been critical evaluation of its use in particular cases by reviewers, editors, scientists collaborating on publications, and other scientists.[24]
   c. Therefore method M probably didn't mislead scientists often; it probably helped scientists learn about aspects of the world that are part of the focus of study in domain D.
2. Argument for the probable approximate truth of assumptions that method M depends upon:
   a. A method that is justified by or depends essentially on an assumption A is unlikely to help scientists learn about the world unless that assumption is at least close to the truth.
   b. Method M is justified by or depends essentially on a particular assumption A about aspects of the world under study.
   c. Therefore assumption A or something similar to it is probably approximately true. (This last point could be refined by distinguishing between aspects of a complex assumption—those whose approximate truth makes a difference to the success of the method and those that don't—but I think the general idea is sufficiently clear; see §2.1 and Potochnik 2017.)

So, if method M is crucial to repeatedly, apparently successful research, and an assumption A must probably be approximately true in order for this to be so, the success of M is evidence for the approximate truth of A.[25] For example, in the preceding section I argued that common, successful practices in which probabilistic computer simulations are assumed to be good models of evolutionary processes (M) would be unlikely to be successful unless those processes realized causal probabilities (A). Clearly such inferences are defeasible: additional information can undermine any of the steps in the argument.

Obvious examples like phlogiston-based studies of combustion and oxidation illustrate this point. Wimsatt (2007) provides other examples.

2.4.2 QUESTIONABLE RESEARCH PRACTICES AND DOUBTS ABOUT *P*-VALUES

I seem to have the misfortune of writing a book containing arguments based on the validity of scientific practices (see above) during a period in which science itself has questioned scientific practices. The concerns that have been raised focus largely on scientific replication, questionable research practices, and certain uses of *p*-values in statistics. In this section, I briefly argue that such concerns, though important, are unlikely to undermine my arguments from scientific practice. (Readers who don't share the relevant worries can skip this section.)

The so-called *replication crisis* became a big topic among scientists and philosophers of science in 2015, when a consortium of researchers attempted to replicate one hundred published psychology studies, with new teams of researchers following the experimental procedures used by the original authors (Open Science Collaboration 2015). Although many of the studies were considered to have been successfully replicated—new results were sufficiently similar to earlier ones—the number that were not replicated was indeed shocking. A small percentage of replication failures would be normal, but if the studies that were not replicated had in fact provided good evidence for their conclusions, it should have been possible to replicate more of them. A common view is that the problem resulted from *questionable research practices* that have the potential to produce scientific evidence that is misleading at best, even when well intentioned. There are reasons to think that such practices are not uncommon in evolutionary biology (Fraser et al. 2018). Another concern is with journal publication norms that can distort the body of evidence produced by publications (R. Rosenthal 1979; Ioannidis 2005). There are also cases of outright fraud in science: fabrication of data or images, for example (Stroebe, Postmes, and Spears 2012; Bik, Casadevall, and Fang 2016).

Failures of replication, questionable research practices, and research fraud constitute a huge area of research and discussion. I cannot do justice to it here. The issues are relevant to my arguments, though, because if it turned out that, in general, patterns of scientific practices on which my arguments were based were not in fact practices that should be considered successful guides to reality, that would undermine my arguments from scientific practice. On the other hand, the practices on which my arguments depend are so widespread in evolutionary biology that if they were generally invalid, the soundness of my arguments should be the least of our worries. Rather, it might not be worth asking any substantial questions about evolutionary biology.

Below I'll briefly indicate why I think that research practices in evolutionary biology should, in general, be trusted, despite potential problems that need to be addressed. (Some readers may not need to be convinced of this point.) After a brief discussion of fraud, I'll focus on one questionable research practice, "p-hacking" (Nuzzo 2014; Head et al. 2015), since it connects most directly to my arguments from practice. Similar arguments could be given concerning other questionable research practices or journal practices that bias the publication record.

First, there is little reason to think that fraud is common in evolutionary biology, and my arguments would not be undermined by low levels of fraud among evolutionary biology researchers. A meta-analysis of studies of the prevalence of fraud suggested that the percentage of scientists who engage in fraud is in the low single digits, with a higher percentage among medical researchers (Fanelli 2009). Fanelli suggests that fraud may be more common in medical research because of financial rewards for medical research. This kind of temptation would be rare in evolutionary biology.

On the other hand, p-hacking is among the questionable practices that seem to be too common in evolutionary biology (Fraser et al. 2018).[26] Given a single experimental setup, a scientist can often keep collecting data or trying different statistical tests until they get a statistical result that would traditionally be considered to provide evidence for a result—that is, until they get a sufficiently low p-value (§2.1). If a scientist then publishes a report of only the experiments and statistics that seem to provide evidence, without noting that their original experimental design failed to produce sufficient evidence, that's considered p-hacking. This is loosely analogous to claiming that your coin is biased toward heads because you kept repeating the same five-toss experiment until you got a five-head sequence—let's say, after forty tries. You would eventually find your five-head sequence, almost surely (note 26, §0.3.4), even if the coin was quite biased against heads. Similarly, p-hacking depends on the fact that experiments and tests can produce "good" results just by chance, even when the property measured is nonexistent. The term "p-hacking" is also sometimes used for the practice of rounding down a p-value that is close to a desired cutoff value, such as 0.05 (Fraser et al. 2018).[27] Thus when I argue from widespread practices in successful empirical research (in particular, practices in which statistical methods are central), the prevalence of p-hacking raises the possibility that these practices should not in fact be considered successful.

However, the available evidence doesn't suggest that research in evolutionary biology is largely untrustworthy. It's relevant that even in the big psychology replication study that raised alarm bells (Open Science Collaboration 2015), most of the results were replicated. Fraser et al.'s (2018) survey of researchers in ecology and evolutionary biology did not show that most researchers engage in questionable research practices. Head et al. (2015) performed a statistical meta-analysis of reported p-value patterns in open-access publications in the PubMed database, which covers a variety of biological fields including evolutionary biology. The authors argued that the distribution of p-values suggests that although p-hacking is too common, and there are probably individual studies whose results are invalid because of it, the pattern in p-values suggest that p-hacking has not resulted in an overall invalidity of scientific results.

Further, empirical results in evolutionary biology show a great deal of convergence and harmony across different research methods and subject populations, when viewed from an appropriate perspective. This seems unlikely if empirical results were largely untethered from an underlying reality. This point is somewhat subtle and could justify a book of its own. The problem is that evolutionary biology is not like fundamental physics, where there are precise theories that constrain a great deal of research. However, it's also unlike psychology or neuroscience, where some areas of investigation can seem largely isolated from others. So let me explain what I mean.

First, note that most empirical studies in evolutionary biology cannot be replicated in any direct manner. Field studies, such as ones I discuss in some parts of this book, sample from natural populations that continue to evolve and whose environments undergo change. Thus, coming back to the same population later to perform the "same" study may find a population that is in a very different state. Even experimental studies, such as the fruit fly study discussed above or other studies I'll describe, typically face a similar difficulty with replication. Although environmental conditions for experimental populations can be controlled, the populations in a "replication" would be different, either because the organisms are sampled from a natural population that has been evolving or because they come from a laboratory population that has itself been evolving. (A notable exception comes from experimental research on microorganisms, such as the studies of *E. coli* evolution by Lenski and his colleagues; see §5.3.3 and §7.3.2.)

Nevertheless, there is a kind of "replication" at a more abstract, theoretical level. There is a great deal of overlapping, convergent evidence within evolutionary biology, and between it and related fields. Genetically related species show commonalities and differences that are consistent with evolutionary processes, patterns in the genome are consistent with evolutionary processes that can be investigated in other ways, and in general scientists repeatedly learn about the same or related properties through genetic, physiological, behavioral, and other studies. For example, there are patterns that are common in particular phyla, such as roles of carotenoids in coloring and parasite resistance among birds or in certain fish (Olson and Owens 1998). Evidence of selection on particular genes can come from statistical analyses of the genome, as well as studies of gene expression and the roles of particular genes in adaptation (e.g., Voight et al. 2006). Moreover, even though particular studies can't usually be replicated in the traditional sense, studies of genetically related populations (see chapters 5, 6) are studies of populations with known sorts of probabilistic relationships. The same is true of studies of the same population at different times. The substantial degree of multifaceted harmony in patterns of information that come from evolutionary biology and related fields, such as genetics, molecular biology, ecology, agriculture, and health research, seems unlikely if most research in evolutionary biology didn't at least provide probable, approximately correct conclusions about the world.

On another point, I want to acknowledge that there are philosophers of science, statisticians, and scientists who seem to shudder when p-values are mentioned. Some of their distaste for p-values probably comes from the ability of scientists to abuse p-values (as in p-hacking). That's not a reason to reject reasonable uses of p-values, though.[28] Other distaste probably results from ongoing debates about statistical methodologies. P-values arise only in frequentist statistics, are ubiquitous in its applications, and are closely tied to assumptions that critics of frequentist statistics reject.

Regardless, it seems clear that frequentist statistics—and p-values—are pervasive in empirical research in evolutionary biology.[29] If this is correct, then I don't see how we can treat that research as successful and as a basis for inference about the world but deny that p-values' crucial role in research is part of its success. I don't thereby reject criticism of uses of p-values and calls for improvement, nor do I deny that an alternative such as Bayesian statistics might possibly provide a better foundation for science. What I believe is that current and past evolutionary biology, with its use of p-values and frequentist statistics, has been quite good enough, enough of the time, that we can start from the assumption that p-values help to provide evidence for conclusions in empirical research. I believe that those who agree that empirical research in evolutionary biology is generally of value but who reject the validity of its typical statistical methods ultimately have a burden of proof to provide an explanation of how so much good science has been done with such supposedly bad scientific methods. As far as I know, this kind of explanation has not been provided. So I ask that p-value skeptics put aside their discomfort while reading the rest of this book.

The arguments in this section have been quite condensed. I invite readers who are not convinced at this point that empirical research in evolutionary biology is, in general, probably a useful guide to approximate conclusions concerning its subjects (see §2.1) to view my arguments from practice as having conditional form: *if* empirical research in evolutionary biology of such and such sorts is generally successful, *then* such and such follows.

2.5 Conclusion

In this chapter, I began by sketching the idea that modeling and statistics in evolutionary biology provide incomplete evidence for causal patterns (§2.1). I then introduced the concept of causal probability in section 2.2. Section 2.3 argued that certain ways that models are used in empirical research show that probabilities in population-environment systems are causal probabilities. This argument provided an illustration of the kind of argument from empirical practice that I discussed in section 2.4. These core ideas, along with those discussed in chapter 1, will come up repeatedly in the rest of the book.

Although I've argued that population-environment systems, if they exist, realize causal probabilities, I haven't properly argued for the idea that evolution should be understood as taking place in population-environment systems. Clarifying and motivating that claim is the task of the rest of the book.

3

Irrelevance of Fitness as a
Causal Property of Token Organisms

3.1 Introduction

A well-known evolutionary biologist, Stephen Stearns, once described fitness as "something everyone understands but no one can define precisely" (Stearns 1976, 4). Stearns knew, though, that he and other biologists do often define fitness quite precisely (e.g., Stearns 1992), albeit in multiple, incompatible ways. The practice of using incompatible fitness concepts has been the basis of much insight about evolution, both through empirical research and theoretical modeling. One point of this book is to clarify this practice by sorting out different roles that fitness concepts play. To that end, I introduce a classification of fitness roles in this chapter.[1] Before I go into detail about the structure of relationships between these fitness roles, some ground has to be cleared, and a foundation for the relationships has to be laid. The present chapter focuses on a ground-clearing task. I explain why an appealing and apparently sensible kind of fitness concept—one that has been the focus of a great deal of valuable work in philosophy of biology—is in fact largely irrelevant to evolution. The view that evolution depends fundamentally on this sort of *causal token-organism fitness* is an impediment to a deeper understanding of evolutionary biology—which almost never uses this idea. It is other fitness concepts that matter for evolution. I begin to develop that view in chapter 4, but I don't fully elaborate it until chapter 7, after additional ground-clearing and foundational work. The result will be a new picture of the nature of evolutionary processes.

In section 3.2, I introduce basic assumptions that will help to organize that picture, and I give my classification of fitness concepts. In section 3.3, I present a series of arguments against the idea that causal token-organism fitness is useful for understanding evolution. In section 3.4, I argue against the view that certain experiments on plants justify the traditional concept of the propensity interpretation of fitness, which is best understood as a kind of causal token fitness. Finally, the appendix summarizes basic mathematical facts about the Price equation, which is mentioned in some places in the chapter and will come up again in chapter 7.

3.2 Kinds of Fitness Concepts

3.2.1 INITIAL ASSUMPTIONS

To give us a place to start, I'll make the following provisional assumptions, some of which preview claims I'll argue for later in the book. I find the assumptions to be more or less plausible, but I know that some readers will not. I hope those readers will ultimately be convinced by the way that the assumptions fit into the picture of practice in evolutionary biology that I develop in this book.

1. Natural selection occurs when the frequencies of inheritable (§0.3.5) types (alleles, genotypes, phenotypes) in a population change over time *because* these types have different fitnesses, or when frequencies remain the same *because* the types' fitnesses are the same. This "because" is a causal "because." I believe that this captures a core intuition about the concept of natural selection.[2]

2. Causal conceptions of fitness must—at least—allow the possibility of natural selection over generational time. Natural selection, and therefore fitness differences, must be capable of causing properties of distributions of organisms and traits in the world. This means that natural selection must in principle be able to act, at least sometimes, in a sustained manner over many generations. This generally means that if fitnesses change, they must do so slowly or in systematic ways.

3. Fitness, in any causal sense of the term, can be defined partly in terms of probabilities causally relevant to the number of descendants that instances of a biological type have. These probabilities should be ones that could be understood as causal probabilities, on the assumption that manipulating the probabilities should also tend to manipulate frequencies of outcomes (§2.2).

4. An environment of a population can be viewed as corresponding to a range of variation in conditions that might be experienced by members of the population over a specified interval of time, along with probabilities of such conditions being experienced (see chapters 1, 2, and most of the chapters after this one).

5. In studying evolution, biologists make some choices, at least in a rough sense, about what aspects of the world to investigate. In particular, biolo-

gists choose ways to delineate populations and what properties of populations to study. This last choice includes a choice about an interval of time
over which the population might evolve, and with it, what environmental
factors might be relevant to that evolution. It would be theoretically convenient to restrict populations to sets of organisms experiencing a relatively
uniform environment, and to sets of organisms unlikely to experience gene
flow to or from other populations. However, those restrictions would mean
that many clear cases of evolution wouldn't count as evolution, or that it
would not be possible to study such cases. Neither restriction is observed
in actual biological practice, I believe. What is required is just that the possibility of subpopulations experiencing different conditions and the possibility of gene flow in and out of the population be taken into account,
either by incorporating these possibilities into models and measurements
or by having reasons to think that their effects can be ignored. (I provide
arguments for these claims in chapters 5 and 6.)

6. Nevertheless, for a specified population in a specified environment in a
   particular period of time, natural selection takes place independent of our
   decisions, modeling, or empirical studies. (See chapters 5, 6, and 7.)

3.2.2 FITNESS CONCEPTS

In evolutionary biology, "fitness" is also expressed by other terms (see §0.3.2
for a few illustrations), as well as by parameters and variables in models.
Some fitness terms are defined in multiple ways, but the ambiguities don't
usually cause problems for researchers. Context, area-specific traditions, and
researchers' explicit definitions make intended senses clear enough for practical use. As suggested in section 0.3.2, I'll use "fitness" as a blanket term for
all concepts intended to capture some general idea about how differences in
traits contribute to natural selection. For my purposes, it will do no harm to
refer to all such concepts, which in some way quantify differences in actual or
probable contribution to the composition of a population at a later time, with
the same term: "fitness."

I do want to make a distinction between different sorts of fitness concepts
or roles, though. I'll start by distinguishing between concepts or roles that are
essentially tied to methods of empirical measurement, as opposed to those
designed to capture causal factors that influence evolutionary outcomes.[3] In
evolutionary biology, as opposed to philosophy of biology, I think it's rare
that these two kinds of ideas are mixed in one fitness concept, but chapter 7
includes some illustrations of hybrid concepts. I'll begin by applying the mea-
surement/causal-factor distinction to fitness treated as a property of a token
organism, and then apply it to organism types.

TABLE 3.1: Classes of fitness concepts defined in the text

	Readily measurable	Not readily measurable
Token	Measurable token-organism fitness (measurable token fitness)	Causal token-organism fitness (causal token fitness)
Type	Statistical organism-type fitness (statistical type fitness)	Causal organism-type fitness (causal type fitness)

Just mathematical
Purely mathematical fitness

Measurable token-organism fitness. A *measurable token-organism fitness* concept, or, briefly, a *measurable token fitness* concept, is one that allows measuring fitness as a property of a token individual, where that property is thought to be relevant to evolutionary success; see table 3.1. (I intend "evolutionary success" to be vague, capturing the idea of increase in frequency in future generations or at least maintenance of a type in the population; this vague notion will be sufficient for the present.) Measurable token-organism fitnesses should be fairly concrete. For example, in later chapters I'll discuss Byars et al.'s (2010) measurement of the total number of children of each woman in the Framingham Heart Study by her number of live births before menopause. This is a measurable token fitness concept.[4] A measurable token fitness concept need make no explicit reference to anything directly involving survival, reproduction, or descendants. Morris, Lenski, and Zinser's (2012) definition of fitness in terms of the energetic or resource-usage benefits of a loss-of-function mutation is one example. Measurements of seed mass (Dong, van Kleunen, and Yu 2018; Hamann, Weis, and Franks 2018) is another. These are measurable token fitness concepts, because the point of measuring such properties is to get evidence for differences between token organisms that are potentially relevant to natural selection.

Causal token-organism fitness. A *causal token-organism fitness*, or *causal token fitness*, concept, on the other hand, attempts to capture the idea that a particular individual in its particular circumstances (see below) has one or more tendencies to realize properties relevant to evolutionary success. Such concepts differ from measurable token fitness concepts, in that causal token fitness concepts are intended to capture some underlying causal character of a token organism in relation to its environment. I described a few such concepts in section 0.3.2, and I will mention others below. For now, I'll note that central versions of the propensity interpretation of fitness (PIF) describe causal token fitness concepts (e.g., Beatty and Finsen 1989; Brandon 1990, chap. 1; Ramsey 2006). Matthen and Ariew's (2002) concept of vernacular fitness seems to be a kind of causal token fitness (see also Walsh, Ariew, and Matthen 2017; Pigliucci and Kaplan 2006). Walsh, Ariew, and Matthen (2017) make it clear that "vernacular fitness" is intended to refer to token-organism fitness concepts that are common among advocates of the PIF.[5] Bouchard and Rosenberg's (2004) characterization of "ecological fitness" in terms of an organism's ability to solve design problems can also be viewed as a causal token fitness concept. Earnshaw (2018) defines a concept of "individual fitness" that provides another example. Next, I distinguish between two classes of fitness concepts that apply to types.

Statistical organism-type fitness. A *statistical organism-type fitness*, or *statistical type fitness*, concept is one that defines fitness as a property of an inheritable type, in such a way that fitness becomes a mathematical function of measurable token fitnesses. For example, if we define the fitness of a trait as an average of the number of offspring that actual individuals with that trait have in a certain generation—that is, if we add up the numbers of actual offspring and divide by the number of parents—then we are treating the trait's fitness as a statistical type fitness. Stearns (1976) and de Jong (1994) surveyed a variety of statistical type fitness concepts in use by biologists. Pigliucci and Kaplan's (2006) "formal fitness" seems to be a statistical type fitness concept. Walsh, Ariew, and Matthen's concept of "trait fitness" (2017), or "predictive fitness" in Matthen and Ariew (2002), plausibly are statistical type fitness concepts, at least in some contexts, but some of these authors' arguments may treat these fitness concepts as purely mathematical fitness concepts (see below).

Causal organism-type fitness. A *causal organism-type fitness*, or *causal type fitness*, concept, by contrast, is one that treats the fitness of a trait as an underlying property that could affect the trait's evolutionary success. As noted in section 0.3.2, PIF advocates often define causal type fitnesses as an average of causal token fitnesses. Chapters 4 and 7 will describe what I see as better ways to think about causal type fitness.

Note that by using the word "measurable" in the name of a category of fitness concept, I don't imply that the other sorts of fitness are by definition *not* measurable. Fitnesses of all of the above sorts are measurable, at least in principle, but some are more difficult to measure than others. Causal token fitnesses such as those that play a fundamental role in the PIF are theoretical quantities and would only be measurable by very indirect means (if they can be measured at all).[6] The same will turn out to be true of causal type fitnesses. Similarly, by using "causal" in the names of causal token-organism fitness and causal organism-type fitness, I don't mean to imply that other sorts of fitnesses can't have effects. For example, having more or fewer offspring (variation in a kind of measurable token fitness) certainly has effects! The point is that the roles of causal token fitness and causal type fitness in evolutionary biology or evolution are supposed to be that they capture facts about causal influences that might contribute to evolution by natural selection, without regard to ease of measurement per se.

Purely mathematical fitness. Finally, a *purely mathematical fitness* concept is a mathematical concept, defined for use in certain mathematical models, that might be usefully be interpreted as one of the other kinds of fitnesses in particular research contexts. Consider the Price equation, a mathematical relationship that's important in evolutionary biology. De Jong (1994) seemed to treat fitnesses in Price's equation as fitnesses of traits, whereas Price (1970, 1972) himself seemed to treat the fitnesses in his model as fitnesses of token organisms. The idea is that the Price equation itself doesn't determine whether the fitnesses mentioned in it are properties of types or tokens, as long as a particular application doesn't force the issue. (I spell this point out in detail in an appendix to this chapter, §3.6.) I want to allow for the fact that mathematical fitness concepts can be used without a particular interpretation, but this conceptual role will do little work in the rest of the book.[7]

Because a great deal of philosophical discussions of fitness concepts have assigned causal token fitness an important role in thought about evolution, I devote most of the rest of the chapter to criticisms of this conception of fitness and ideas associated with it. Even those unfamiliar with philosophical debates about fitness (§0.3.2) will, I hope, find the rest of this chapter helpful for understanding later discussions. I want to emphasize that in criticizing the utility of causal token fitness concepts here and in later chapters, I recognize that measurements of fitnesses of token individuals play important roles in evolutionary biology. Such fitnesses are *measurable* token fitnesses on my view and not causal token fitnesses. I discuss them in chapter 7.

3.3 Why Causal Token Fitness Is Problematic

As I've mentioned, the idea of fitness as causal property of particular token organisms seems to be popular among philosophers of biology (e.g., Brandon 1978, 1990; Mills and Beatty 1979; Bouchard and Rosenberg 2004; Ramsey 2006; Pence and Ramsey 2013). It's taken by many to be at the core of causal conceptions of natural selection, and it has played various roles in attacks on causal conceptions of natural selection (e.g., Matthen and Ariew 2002, 2009; Walsh 2007, 2010; Ariew and Ernst 2009; Walsh, Ariew, and Matthen 2017). The rest of this section will present a series of arguments against the significance of causal token fitness for evolutionary biology and evolution.[8] It might be unfair to attribute all of the assumptions on which these arguments are based to every author who argues for causal token fitness concepts or takes their significance for granted. The problem is that it's not always clear what assumptions authors do make about the issues I discuss below. My arguments should at least show what ways of thinking about causal token fitness are problematic.

3.3.1 LEWONTIN'S CONDITIONS FOR NATURAL SELECTION

I'll start with a general characterization of natural selection that I believe captures one common theoretical role of fitness in evolutionary biology. In particular, I'll take as a starting point Lewontin's formulation of the conditions required for natural selection by Darwin. Lewontin wrote:

> As seen by present-day evolutionists, Darwin's scheme embodies three principles . . . :
>
> 1. Different individuals in a population have different morphologies, physiologies, and behaviors (phenotypic variation).
> 2. Different phenotypes have different rates of survival and reproduction in different environments (differential fitness).
> 3. There is a correlation between parents and offspring in the contribution of each to future generations (fitness is heritable). (Lewontin 1970)

Lewontin's formulation spells out paradigmatic conditions for natural selection in a perspicuous manner, and similar formulations have been given by others (e.g., Godfrey-Smith 2009b). Lewontin suggests that these three conditions are necessary and sufficient for evolutionary change by natural selection. In fact, they are neither necessary nor sufficient for natural selection (Godfrey-Smith 2007, 2009b). However, they capture the core of the notion of evolution by natural selection sufficiently well for my purposes here, and they provide a starting point for my discussion.

3.3.2 RECOMBINATION

Warren Ewens, a well-known population geneticist, wrote:

> First, while it is universally agreed that fitness is a property of the entire genome of an individual, it is also apparently agreed, with Wright (1931), that to a first approximation, for a short time, a constant net selection value of any allele may usefully be defined. (Ewens 2004, 277)

I think that Ewens's point is probably that interactions between genes can affect fitness, and so, in theory, any part of the genome could be relevant to fitness. However, if we treat fitness differences as the basis of natural selection, fitness treated as fundamentally a property of an entire genome won't work. It's worth noting that Ewens's remark occurs in a book that surveys a broad range of population genetics models; I don't think that any of the models treats fitness as a property of an entire genome (though there are a lot of models in the book, and I may have missed something). This is as it should be, I believe.

If fitness was attributed to an entire genome, the inheritability of fitness across more than a few generations would often be very low in species that undergo significant recombination (§0.3.5). Recombination occurs in many sexually reproducing species, and it results in pieces of DNA from an individual's two parents becoming interspersed in the individual's chromosomes. Subsequently, when that individual produces offspring with the help of a mate, the individual's DNA will be interspersed with that of the mate's. This process continues down through the generations, gradually mixing together different patterns of genetic variants. So if natural selection were to be understood, in general, as the result of differences in whole-genome fitness, it's difficult to see how it could act in a sustained way over many generations, as required in section 3.2.1. This is not to say that natural selection can't act over the short term, but a general concept of fitness shouldn't be defined so that it would be difficult for it to be able act over the long term. Of course, in species without recombination, including many asexual species, we could treat fitness as a property of the genome, but this is no longer a claim about a general concept of fitness. Notice that causal token fitness attributes fitness to the organism as a whole, which includes the genome, so the same point applies: causal token fitness per se can't play a sustained role over many generations in species with recombination. (Some will object here that causal token fitness is often thought to contribute to causal type fitness, which *can* play a sustained role over generations. I explain below why the idea that causal type fitness depends on causal token fitness is problematic.)

The preceding argument is related to one that George Williams gave in his *Adaptation and Natural Selection* (1966), which was subsequently popularized by Richard Dawkins in *The Selfish Gene* (1976).[9] The argument is usually described as implying that only very small segments of DNA can be units of selection. That is, it is patterns in small segments of DNA (idiosyncratically called "alleles" by Williams and Dawkins) that are selected for or against—rather than organisms or phenotypic traits or some other entities. Williams and Dawkins argued that since, over the course of generations, recombination eventually breaks up all long segments of DNA (with high probability), only small segments of DNA persist long enough to have a significant effect via natural selection. It's important to notice that in this account, alleles function as types that are realized by individual organisms. This can be seen from the fact that Williams and Dawkins measure the effects of natural selection by counting token organisms that bear particular alleles, rather than, for example, counting all of the tokens of a given allele that might be found in an organism's cells (Sterelny and Kitcher 1988). Thus what Williams and Dawkins argued for was that natural selection only acts on the distribution of organism types defined by what they called "alleles." Williams and Dawkins's position is too extreme, however. It's not necessary that an inheritable trait such as an "allele" persist for many thousands of generations. All that's necessary is that it can persist long enough for natural selection to have an effect (see §4.3). The problem with attributing fitness to an entire genome is that it means that there will be many cases in which that's not possible. Selection requires that fitness-affecting variation be able to be passed on, as Williams argued. It does not require that it be passed on perfectly, as Lewontin's statement of conditions for natural selection make clear.[10]

3.3.3 CIRCUMSTANCE VARIATION

There is a related, deeper problem with causal token fitness if it is defined so as to depend not only on an individual's genome but also on the way in which the genes interact with particular environmental circumstances, either during development or over the course of mature life stages.[11] The PIF seems to have the implication that causal token fitness depends on all of an individual's particular environmental context, but authors are not always clear about this point.[12] However, if fitness is a property of a token organism, I don't see how to avoid the assumption that any variation in its particular environmental circumstances might affect its fitness. What I mean here is that variations that can affect the reproductive success of an individual are not limited to the kinds of patterns explicitly referenced in models and empirical studies of environmental variation (often labeled with terms such as "patch," "habitat," "niche," and "environment"). Rather, what might be called "microhabitats," or what I call "circumstances" (Abrams 2007a) can matter to token fitness as well. These include any particular, detailed configuration of properties of the environment that could affect probabilities of reproductive success for a token organism, as I'll explain.

The extent to which a token organism's fitness depends on minute environmental variation might be clear to some readers from the description in section 1.2, but it's worth giving more specific illustrations. Suppose that while searching for a particular kind of nest site, a young male house sparrow happens to fly below a small opening in the forest cover at the moment a hawk happens to fly overhead. Consequently, the hawk sees the sparrow, chases it, and injures it but doesn't manage to kill the sparrow, which escapes. Because of the injury, the sparrow gets an infection when it washes its feathers in a pond, which happens at that time to contain bacteria to which the house sparrow has low resistance. The house sparrow's resistance is low, in turn, because its diet has been limited to a few foods; this is in part a consequence of a recent landslide that caused rainfall to be diverted to other areas. The infection leads to the house sparrow's subsequent death after it manages to mate once. As a result, it has two offspring—fewer than it might have had if it had managed to live longer. One can easily imagine, though, that if any of the circumstances I described had been a little bit different, the house sparrow might have been likely to have more, or fewer, offspring. According to the PIF, causal token fitness has to do with probabilities of a token organism having different numbers of offspring. The point here is that

1. natural variation in circumstances should make some outcomes *extremely probable*, and others *extremely improbable*; and
2. small variations in circumstances can make a huge difference in *which* outcomes are probable.

There's no obvious limit to the sorts of minute variations that might affect such probabilities. For example, the fitness of a prey might be affected by fact that a leaf is blown in such a way as to allow the prey to be noticed by a predator, simply because the movement of the leaf, from a particular angle, was similar to certain movements of potential mates to which the predator's cognitive system is particularly sensitive. On the other hand, a simultaneous shift in the wind might mean that the predator becomes distracted by the smell of a conspecific before noticing the prey. My view (Abrams 2007a), in fact, is that particular organisms' interactions with their environments might as well be deterministic (see chapters 8, 9), in which case the PIF would make causal token fitnesses equal to actual numbers of offspring, actual numbers of descendants over some period, or some function of actual descendants. However, even if one thinks there is always some significant objective probability that actual outcomes for a token organism could have been different, it still appears that most outcomes would have extremely low probabilities, with only narrow ranges of outcomes having high probabilities.

However, any such fitness-determining circumstances for a token organism would not give it a high or low fitness in any sense relevant to natural selection. For a token organism's particular set of circumstances will almost certainly never recur. Any relative success that the organism has in responding to its *specific* circumstances as such can't be passed on to offspring—since the offspring will not experience the same or even very similar circumstances. An organism's particular set of circumstances is therefore not the sort of thing to which natural selection can respond. Of course, that hawks are present in the region and are sometimes overhead matters to natural selection. That injuries of various sorts occur matters to natural selection. That infections of various sorts occur matters, and that resistance to them can be low matters. That the water supply to nearby plants is sometimes low matters. These things matter because they have a significant probability of recurring. What matters to natural selection is not this or that organism's particular circumstances and particular fate but the sorts of conditions that individuals in the population are likely to encounter repeatedly.[13] (See chapter 4 for more on these points.)

This point extends to Ramsey's (2006) view that causal token fitness depends on a *fitness environment*, which is defined in terms of probable conditions that descendants of an individual might encounter over the course of many generations. Since Ramsey defines a fitness environment to be relative to a particular individual, different individuals in the same population can have different fitness environments, even if they are genotypically and phenotypically identical. Although fitness environments of different individuals might overlap as their descendants interact and share space, each descendant still faces its own particular circumstances. The same point applies to Pence and Ramsey's (2013) definition of fitness in terms of a particular way of averaging a token organism's descendants' probable reproductive successes.[14]

Now, it's reasonable to think that in most species there will have been selection for robustness of patterns of survival, reproduction, and so on in the face of environmental variation (A. Wagner 2005; see Wimsatt [2007] for related discussion of robustness). After all, it's well known that greater variation in numbers of offspring often counts as a reduction in number of later descendants (§0.3.2). Thus robustness to environmental variation—reducing variation in reproductive outcomes—will often be selected for. One might argue, then, that because of selection for robustness, variation in environmental circumstances usually won't matter to fitness; organisms in a population that were phenotypically similar would have similar token fitnesses, regardless of what circumstances they experienced within the range of variation that is likely for a given environment.

However, complete robustness to environmental circumstances is unlikely to occur. That would mean, implausibly, that a given type of organism would have the same number of descendants for any set of circumstances included in the range likely for a population-environment system. Moreover, a population with a great deal of robustness to environmental variation is unlikely to persist in that state for very long. It's likely that fitter variants would eventually arise that allowed the exploitation of new, varying resources, despite greater risk of failure; the long-term cost of variance in number of offspring can be outweighed by the benefit of an increased average number of offspring ($0.3.2). In the end, natural selection favors more descendants overall, even if they come at the cost of reproductively unsuccessful descendants who suffer from variation in circumstances along the way. Further, it may be that a great deal of robustness to environmental circumstances is difficult to achieve, either because a population lacks the genetic variation that would allow for selection for robustness to many different circumstances, or because robustness of some sorts requires too much energy, or because robustness incurs other costs that lead to selection against it. Thus, even though there is probably a great deal of selection for robustness, it's plausible that there is also a great deal of dependence of causal token fitness on circumstance variation—as illustrated by examples above. So it seems likely that causal token fitness will only rarely, if ever, be sufficiently inheritable to play a direct role in natural selection.

3.3.4 ARBITRARY FLUCTUATIONS IN TRAIT FITNESS

Advocates of the idea that causal token fitness is the fundamental kind of fitness often adopt the view that the fitnesses of a trait should be defined in terms of a simple average of the fitnesses of the actual token organisms that have that trait in a population at a given point in time, or during a period of time (Brandon 1978; Mills and Beatty 1979; Sober 1984, 2000; Bouchard and Rosenberg 2004; A. Rosenberg and Bouchard 2005).[15] Since advocates of this view are usually causalists, who presumably think that differences in trait fitness can cause evolutionary change, I'll treat these trait fitnesses as causal type fitnesses.

The problem with this proposal is that it means that causal type fitnesses can fluctuate in odd ways from one time period to the next, as individuals in the population happen to experience this or that "lucky" or "unlucky" set of microenvironmental circumstances.[16] Suppose that in one year, in one population of birds, many birds with a particular trait—a tendency to seek out certain kinds of nest sites, for example—experience similarly unlucky circumstances, while the following year, very few birds with the trait do so. If we say that causal token fitnesses can depend on circumstances that are encountered, and that the causal type fitness of the nest-site-affecting trait is the average of those causal token fitnesses, we have to say that in the first year the trait had a low fitness, while in the second year it had a high fitness. However, in the situation as described, this is just a difference in chance encounters due to particular circumstances as each bird wends its way through the environment. The overall probabilities of environmental variation in the population need not have changed at all, and it might be very strange for a biologist to assume that they had. The fact that the trait "had a bad year" or a "good year" shouldn't necessarily reflect its overall probability of success in the population over a longer term. (Similarly, the fact that a gambler at a casino has a good or bad run while repeatedly betting on a single outcome doesn't mean that the chance of the outcome keeps changing.)

One can, of course, take the average of causal token fitnesses over several years, or over a larger population (see chapter 5). Assuming that the probabilities associated with the population-environment system don't change, this would make a run of many unlucky or many lucky encounters over multiyear, large-population collections of organisms improbable. That is, it would be improbable for causal type fitnesses, defined as averages of causal token fitnesses, to exhibit weird values that didn't correspond to probabilities of outcomes for a trait. However, even if this possibility is unlikely, it's still a problem. If fitness is part of the fundamental nature of natural selection, shouldn't it *always* reflect probabilities of evolutionary outcomes? Treating causal type fitness as an average of causal token fitnesses of actual organisms doesn't do this. Of course, one might still want to say that averages of causal token fitnesses are a good way to *estimate* causal type fitnesses. However, that would mean that causal type fitness was something other than an average of actual causal token fitnesses, since we would be using the average to estimate some other thing (§7.4). Also notice that given the traditional PIF intuitions about causal token fitness as depending on nonextreme chances (i.e., ones far from 0 or 1) of numbers of offspring, causal token fitness is not the sort of thing that could be measured in practice, so there would be no basis for the estimate.

Notice that the arguments in this section don't depend on the fact that causal type fitness is being defined as an average per se. The same problem would exist, for example, if we defined causal type fitness as the average of all causal token fitnesses in a given year, minus, say, its variance over some set of surrounding years, where each year is weighted equally. The problem results from defining causal type fitness in terms of causal token fitnesses of actual organisms.

3.3.5 GENERALIZED HITCHHIKING

Sober (2000, 2013) gave another reason not to define type fitness as an average of actual token fitnesses.[17] Here is an illustration. Suppose that there are two pairs $\{A_1, A_2\}$ and $\{B_1, B_2\}$ of heritable traits present in a population. The traits in each pair are mutually exclusive: an organism can have A_1 or A_2 but not both, and similarly for B_1 versus B_2. Each pair might be controlled by alternative alleles at a locus, but that assumption isn't required. For house sparrows, A_1 and A_2 might be two different beak shapes, or variations in the physiological processes that process chemicals such as melanin that help feather color and parasite resistance. And B_1 and B_2 might be variations in vocal cord muscles that affect the pitch of sparrows' calls. Now suppose that, as it happens, in one population, all sparrows with A_1 have B_1, and all sparrows with A_2 have B_2. (There are various ways that this might have come about, but that won't matter here.)

If causal token fitnesses are defined in terms of probabilities of events such as having offspring, and the fitness of a trait is the average of token fitnesses for all actual sparrows with that trait, then A_1 and B_1 must have the same fitness, since exactly the same token fitnesses are averaged over to compute the fitnesses of A_1 and B_1. The same argument shows that A_2 and B_2 must have the same fitness. Yet, intuitively, it seems possible that the B_1 versus B_2 variation in call pitch makes no difference at all to any house sparrow's ability to survive and reproduce, while the difference between A_1 and A_2 makes an enormous difference (or vice versa). In Sober's (2013) terminology, if A_1 is fitter than A_2 in this scenario, then there would be *selection-for* A_1 over A_2, but only *selection-of* B_1 over B_2.[18] Defining fitness as an actual average would mean that B_1 and B_2 had very different fitnesses, even though they in fact make no difference to survival and reproduction; they are merely subject to selection-of.[19]

If we restrict our focus to genes rather than heritable traits more generally, this is just an extreme case of what is known as genetic *hitchhiking* due to linkage disequilibrium (e.g., Gillespie 2004; Rice 2004; Charlesworth and Charlesworth 2010; Lynch and Walsh 2018). If A and B are two loci, we say that their alleles have a high level of *linkage disequilibrium* when the frequencies of, for example, B_1 among A_1 individuals are greater or less than would be expected if the alleles had been randomly combined. If A_1 and B_1 were paired randomly, then on average the frequency of B_1 among A_1 individuals would be equal to the frequency of B_1 in the whole population. In such cases, we say that the loci are in linkage *equilibrium*. There is linkage disequilibrium between the A and B loci when the (actual) frequency of B_1 among A_1 individuals differs from the overall frequency of B_1 in the population.[20]

There is a great deal of research, both theoretical and applied, on the effects of linkage disequilibrium on evolution.[21] A central assumption of this research is that there can be "neutral" alleles that are in disequilibrium with "selected" alleles. Many models then examine ways that fitness differences at a locus undergoing selection can affect the distribution of alleles at a neutral locus that is in linkage disequilibrium with the selected locus. Some of these models are then used in empirical studies. In particular, models show that when there is sufficiently strong selection on one allele A_1, and a possibly neutral allele B_1 at another locus is (empirically) correlated with A_1 in the population, selection on A_1 may cause B_1 to increase in frequency.

In scenarios described by such models, if there is an empirical correlation between A_1 and B_1 within a population, the type-fitness-as-actual-average method implies that the alleles at a neutral locus B are not equally fit. Yet the entire point of these models in evolutionary biology is to examine what happens to the frequencies of B_1 and B_2 when they are—by assumption—equally fit, though correlated with other alleles that are not. (Here B_1, A_2, etc. function as organism types.) I'll give an illustration.

If we let B_i represent either B_1 or B_2, then according to the actual-average definition of type fitness, the fitness of B_i is the average fitness of those A_1s with B_i and those A_2s with B_i. However, the averages computed in this way for B_1 and B_2 need not be equal. For example, suppose that A_1's fitness $w(A_1) = 2$ and A_2's fitness $w(A_2) = 1$, and that 75% of individuals with allele B_1 also have allele A_1, while 25% have A_2. Then the actual-average "fitness" $W(B_1)$ of B_1 would be

$$W(B_1) = 0.75 \times w(A_1) + 0.25 \times w(A_2) = 0.75 \times 2 + 0.25 \times 1 = 1.75.$$

Since the frequency of B_2 among individuals with A_1 is equal to 1 minus the frequency of B_1 among individuals with A_1 (and similarly for A_2), the actual-average fitness of B_2 would be

$$W(B_2) = 0.25 \times w(A_1) + 0.75 \times w(A_2) = 0.25 \times 2 + 0.75 \times 1 = 1.25.$$

Thus, according to the actual-average method of determining causal type fitness, the fitnesses of B_1 and B_2 are different, even though they in fact do not differ in their effects. Not only is this intuitively incorrect, implying that fitness differences and natural selection are not always causes of evolutionary change, but it treats as nonsense an enormous body of work in modeling and empirical research in evolutionary biology. The problem here comes from defining causal type fitness as an average of causal token fitnesses.

Note that there is nothing about the preceding model that's specific to alleles at loci per se. It works just as well with Sober's (2013) example of correlated phenotypic traits. However, I've generalized Sober's trait-hitchhiking argument in a small way by applying it to cases in which correlation between alleles is not so extreme that all A_1s are B_1s and vice versa. Linkage disequilibrium models in fact offer distinctions that go beyond Sober's selection-of/selection-for distinction. For example, the degree to which a selected allele affects a linked allele will typically change over time as linkage disequilibrium evolves, and one can model the effect of linkage disequilibrium between two loci when both loci are independently under selection.

Thus, changes due to fitness as opposed to linkage disequilibrium and hitchhiking are indistinguishable according to the actual-average view of causal type fitness. Moreover, one can't even begin to think about how to address this issue in terms of causal token fitnesses alone, because it depends completely on a population-level property: empirical correlation between traits. There is no way to define linkage disequilibrium without reference to a population.[22]

3.4 Causal Token Fitness in Plant Research?

If there is any case in which causal token fitness could play a role in empirical research like that supposed by traditional versions of the propensity interpretation of fitness, it might be in experimental research on plants. Brandon (1990, chap. 2) can be read as suggesting that causal token fitness in this sense is measured in some plant studies. He discusses experiments in which clones of the same plant are planted in the same soil and then grown under the same conditions (Antonovics, Clay, and Schmitt 1987; Brandon and Antonovics 1996). Since the plants are genetically identical and are grown in the same soil, if they produce different numbers of seeds—as they often do—we may have evidence that there are nonextreme chances of producing different numbers of seeds in identical circumstances. The various numbers of seeds observed then allow us to estimate the causal token fitness of each such plant, since they are, it seems, identical.

However, I see little reason to think that the circumstances (in the above sense) of each seed in such experiments are truly the same. Soil consists of different substances mixed together in a way that is presumably nonuniform, with water flowing past seeds and their roots in different ways. Similarly, plants can experience influences from different airborne substances. In that case, what would be measured by the kind of experiment that Brandon describes is not a causal probability distribution over numbers of seeds based on the overall inheritable type of a seed in one particular environmental circumstance. Rather, what would be measured, on this hypothesis, is variation in soil and air circumstances.[23]

This, in fact, is how some biologists seem to view such experiments: they assume that different outcomes for clones are due to environmental variation. Ordás, Malvar, and Hill (2008) planted different sets of genetically identical maize strains, in what appeared to be the same conditions, in order to measure interactions between genes and what they called microenvironmental conditions:

> We use the term "micro-environment" to refer to the environment of a single plant growing at the same time and nearly at the same place as another plant, specifically that of plants in the same row. (Ordás, Malvar, and Hill 2008, 387)

That is, the scientists assumed that plants growing in nearly the same place were in different microenvironments, with variation in environmental conditions. Note, though, that the scientists did not try to measure microenvironmental differences per se, so it's theoretically possible that the same results could have come from truly identical conditions with indeterministic chance effects on seed development, as Brandon's approach suggests. In that case, the genetic variation that was measured would be variation in sensitivity to indeterministic effects, despite what Ordás, Malvar, and Hill wrote. There seems to be no good reason to reinterpret these researchers' assumptions in that way, though, given arguments above.

Another possibility is that even though the seeds are genetically identical, and even if they were epigenetically identical in the sense that the same cellular context was passed on from a maternal plant, there could be developmental differences between seeds before they were planted. Lewontin (2000) argues that variation in numbers of sensory bristles in *Drosophila* is the result of developmental noise resulting from variation in the ways in which molecules bounce around inside cells. The same kind of phenomenon could affect seed development. However, the stochasticity of such intracellular interaction seems to have to do primarily with complex deterministic effects; whatever quantum mechanical effects are involved are so minute that they are difficult to calculate (Kuriyan, Konforti, and Wemmer 2013, chap. 6). A reasonable way to understand this sort of developmental variation is thus as a kind of deterministic microcircumstance variation (see §8.3.2 and Abrams 2017).

3.5 Conclusion

In this chapter, I introduced distinctions between token fitnesses and type fitnesses, and between different kinds of fitness within each category. Roughly, these finer distinctions are between fitness as something that's relatively easy to measure from observable properties (measurable token fitness, statistical type fitness) and fitness as something that can be considered causal in some sense (causal token fitness, causal type fitness). I discuss these distinctions further in chapters 4 and 7.

Most of the chapter was devoted to criticisms of the usefulness of the idea of causal token fitness (§3.3, §3.4). After providing some initial context (§3.3.1), I argued that views that treat causally relevant fitness differences as properties of actual token organisms have trouble making sense of selection's basic character involving inheritable variation in fitness. First, recombination reduces and potentially destroys the inheritability of token organisms' characters (§3.3.2). Second, variation in environmental circumstances experienced by token organisms can make it difficult for fitness to be inheritable (§3.3.3). Third, this problem cannot be avoided by defining type fitnesses as simple averages of fitnesses of actual token organisms, because it allows type fitnesses to vary in ways that make no sense given the theoretical role of fitness in natural selection (§3.3.4). Fourth, defining type fitness as a simple average conflicts with ways that the concept of fitness is used in linkage disequilibrium models and natural generalizations of them (§3.3.5). Finally, I argued that, despite some suggestions by Brandon, experimental evolution of plant clones does not seem to illustrate a role for causal token fitness in empirical research (§3.4).

Sometimes my arguments depended partly on assumptions listed in section 3.2.1. The first three assumptions there should be generally acceptable, apart from minor details, for most readers sympathetic toward a causalist view. The other three assumptions are part of the general picture that I'll develop in the book. My arguments in this chapter occasionally made use of the fourth assumption, but the fifth and sixth assumptions were listed mainly to avoid complicating the presentation in this chapter and didn't play a role in the arguments above.

I conclude that it's most reasonable to think that fitnesses capable of playing the sort of role needed by the concept of natural selection don't seem to attach to actual token individuals and can't be defined as an average of fitnesses of actual individuals. In later chapters, I argue that type fitness in a causal sense is a theoretical quantity—a property of traits realized in particular population-environment systems—and that this view is what best fits with the concept of evolution by natural selection. Since causal type fitness, in my view, depends essentially on probabilities concerning the operation of a population-environment system as a whole, it can take into account probabilities of changes in both environmental circumstances and effects of recombination. Token fitnesses do have a role to play in the study of causal type fitnesses, but mainly as readily measurable properties—measurable token fitnesses—rather than in the form of causal token fitnesses.[24]

3.6 Appendix: The Price Equation

I mentioned in section 3.2.2 that the Price equation (Price 1970, 1972) provides an illustration of the idea of purely mathematical fitness. The Price equation is a theorem that expresses the change in a population or tendency toward change as the sum of two terms. Part of the interest of the Price equation comes from ways that the terms can be interpreted as representing different factors influencing change in a population (Rice 2004; Okasha 2006; Bourrat 2021).

For the points I want to make about the Price equation, though, it will be sufficient to treat the second term as zero, giving us a simple, earlier version of the Price equation known as the Robertson-Price identity (Robertson 1966; Lynch and Walsh 1998):

(3.1) $$\Delta \bar{z} = \mathrm{cov}(z, w) = \mathrm{E}_i[(z_i - \bar{z})(w_i - \bar{w})].$$

Here z refers to traits and w refers to fitnesses, but "trait," at least, is ambiguous, as I'll explain.[25] In English, equation (3.1) says that

> the change (Δ) in the average value ($\bar{z}$) of a set of competing traits z, from one generation to the next, is equal to the mathematical covariance (cov) of fitnesses w and traits z.

Here we are assuming that traits—heights, for example—can be represented by numbers z_i. More explicitly,

> the change in the average value of a set of competing traits z from one generation to the next is equal to the average (E), for integers $i = 1, i = 2, \ldots, i = n$, of this indented expression:
>
> $(w_i - \bar{w})$, the difference between each of the fitnesses w_i and the average fitness $\bar{w}$,
>
> times
>
> $(z_i - \bar{z})$, the difference between each of the traits z_i and the average trait $\bar{z}$.

Now, what are the w_is and z_is? Well, z_i could refer to a trait of the ith individual (e.g., its height, or perhaps some more complex trait). Then w_i would refer to individual i's token fitness. Alternatively, given a set of traits, or a set of alleles at a locus, z_i could refer to the ith trait or the ith allele. In that case, w_i is a type fitness, since alleles can be viewed as genetic types that can be shared by different individuals. The Robertson-Price identity is true as a mathematical identity either way. Without a particular context that specifies whether we are dealing with token organisms or traits themselves, we should treat fitnesses in the Robertson-Price identity or in the more general Price equation as purely mathematical fitnesses.

4

Roles of Environmental Variation in Selection

4.1 Introduction

Natural selection depends on differences in fitness, and fitness depends on organisms' interactions with their environment. But environments vary in space and time, sometimes in extreme ways. In sections 1.2 and 3.3.3, I focused primarily on very fine-grained environmental variation. More coarse-grained variation, across what biologists call patches, habitats, environments, and so on—or what I'll call *subenvironments* of a *whole environment* over which a population ranges—plays an important role in evolution by natural selection.[1] (Taking the whole environment to be what a population "ranges over" is an informal gloss of the view that I'll present in chapters 5 and 6.) Such coarse-grained environmental variation can help maintain genetic differences within and between populations, increase the chance that new species will be created, select for plasticity (flexibility of individuals' responses to the environment), and select for behaviors that affect the environment with which an organism interacts. Variation in subenvironments can involve both spatial variation and temporal variation, and it can result from habitat choice or migration—since these can expand the variety of environments to which a population is exposed—or from niche construction, in which organisms modify their environments. Relationships between environmental variation and causal organism-type fitness are the focus of this chapter.

I'll discuss simple mathematical models of ways that environmental variation contributes to fitness, and I'll describe a general form of the role of environmental variation and other variations in population-environment systems (§4.2). I'll propose ways to distinguish the role of microenvironmental variation—that is, circumstance variation—that was discussed in chapter 3 from roles of more coarse-grained variation that matter for natural selection (§4.3). I'll also explore ways that causal token fitnesses might be used to define causal type fitnesses in some contexts, taking account of the dependence of causal token fitness on finely varying environmental circumstances (§4.4). By this point, some readers will be wondering when I will finally get around to giving a precise definition of what I see as the correct concept of causal type fitness. Section 4.5 explains why a search for a general definition of causal type fitness for all biological evolutionary contexts is misguided.

In the rest of this chapter, "token fitness" and "type fitness" always mean "causal token fitness" and "causal type fitness," respectively, unless context makes another sense salient.

4.2 How Do Subenvironment-Relative Fitnesses Combine?

Causal type fitnesses—fitnesses of traits A that are inheritable, in the sense that organisms with A will often produce descendants with traits at least similar to A—are not vulnerable to the arguments from recombination or from circumstance variation presented in chapter 3. Unlike whole genomes, such traits can easily be realized many times during the evolution of a population. Moreover, since inheritable traits can be realized repeatedly in organisms that experience different circumstances, the environment relative to which type fitnesses must be defined will be more general than any such circumstances. Questions remain about how type fitnesses relate to environmental variation. The central theme of this section is that for a given partitioning (§0.2.1) of the whole environment into subenvironments E_j, the overall fitness of a trait is a weighted function of fitnesses in each subenvironment, where weights are probabilities that instances of the trait will be realized in each subenvironment.[2]

4.2.1 BRANDON'S THREE CONCEPTS OF ENVIRONMENT

Robert Brandon often seems to treat fitness, or what he calls "adaptedness" (1978, 1990), as causal token fitness, but chapter 2 of Brandon (1990) is relevant to causal type fitness.[3] There he defined three concepts of environment. The *external environment* of a set of organisms consists of any properties of the world external to the organisms. The subset of external environment properties whose variations can affect these organisms' future contribution to the population count as components of the *ecological environment*. Not all variation in the ecological environment is relevant to natural selection between distinct inheritable types, however. For example, if two genotypically different strains of the roundworm *Caenorhabditis briggsae* respond to spatial variation in climate in ways that affect fecundity (Prasad et al. 2011), but the *relative* fitness (§0.3.3) between them is constant, then variation in climate would count as variation in ecological environment but not variation in the *selective environment*. The selective environment varies only when relative fitnesses of competing types vary by subenvironment. Brandon also defined a concept of selective environment "neighborhoods." These are regions of an environment in which fitnesses of competing types don't change very much relative to each other. Note that Brandon often spoke of environmental variation as spatial, using plants as illustrations, but it's natural to extend his notions to complex combinations of variation in many properties over space and time.[4]

Although Brandon (1990, chap. 2) initially proposed that it is selective environments that capture the role of fitness in natural selection, his discussion later in the same chapter shows that factors other than fitness differences relative to selective environments can also matter for natural selection. Here is a schema that illustrates the idea. Suppose that the relative fitness inequalities between two inheritable types A and B are the same in two subenvironments E_1 and E_2; in particular, suppose that

A has greater relative fitness than B in each subenvironment E_1, E_2.

Assume also that

the absolute fitness of both A_1 and B_1 is greater in E_1 than in E_2.

In other words,

both A and B individuals are likely to produce more descendants starting from E_1 than E_2, although A is likely to produce more descendants than B in E_j, whether E_j is either E_1 or E_2.

In this case, the overall fitness of A and B depends on the probabilities of each type being found in E_1 and E_2, and *not* just on their relative fitnesses in each environment (see Glymour [2011, 2014] for similar points). If the difference in absolute fitnesses between E_1 and E_2 is great enough, then B can be fitter than A even if A is fitter than B in each subenvironment. All that's necessary is that B have a sufficiently greater probability than A of landing in E_1.[5]

Brandon's discussion is relevant to my focus in this chapter, but from my point of view, his explicit statements don't go far enough. First, Brandon's own example shows that the concept of a selective environment is relevant to overall selection only in certain cases, since it shows that a type B can be fitter than A overall, even though A is fitter than B in each selective (sub)environment.

That is, B could be fitter in the sense that it has a greater probability of evolutionary success, increased frequency, and so on in the short term or the long term, or both.[6]

Second, Brandon illustrates the case just described using an example of habitat choice, in which insects have inheritable preferences for laying eggs on one kind of plant rather than other. The significance of Brandon's point goes beyond habitat choice, however. For example, if plants have inheritable variation affecting the probability of seeds being blown by the wind, some types of plants in a population may be more likely to encounter certain environmental variants than others, but this would not usually be considered habitat choice. Also note that parts of Brandon's discussion assume that ecological environments do not vary in extreme ways over large portions of a whole environment. However, as I suggested above, and as the house sparrow examples in chapter 1 suggest, it appears that minute variations in circumstances within an environment can produce extreme variations in probabilities of reproductive success.

4.2.2 SIMPLE MODELS

Let's look more closely at some simple models of contributions of environmental variation to causal type fitness. First, suppose that each token organism experiences only a single subenvironment for its entire life, but different organisms experience different subenvironments that affect fitness. For example, suppose that the organisms are plants with wind-blown seeds. Each plant grows in the soil in which it lands, and it experiences a localized pattern of light and shade. Suppose further that which environment a given organism experiences is governed by a single objective probability distribution. Levins (1968) argued that in such cases—of what he called "coarse-grained environments"—the overall fitness should be treated as a probability-weighted average over the m subenvironments. The idea is that each organism of type A experiences *either* subenvironment E_1 with probability $P_A(E_1)$ *or* subenvironment E_2 with probability $P_A(E_2)$, and so on for other subenvironments. However, no organism experiences more than one subenvironment. The evolutionary success of a type A is the sum of its evolutionary success values in different environments, but these should be weighted by the probabilities that A will encounter them. (These probabilities of subenvironments will usually correspond, roughly, to frequencies of encountering them, but they need not; see chapters 1 and 3.) Thus to calculate the overall fitness $w(A)$ of A, we can simply sum fitnesses $w(A|E_j)$ in subenvironments E_j (see note 5)

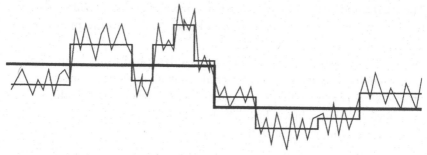

FIGURE 4.1. Schematic representation of variation in fitness in subenvironments at different levels of grain. The horizontal axis represents positions in space, time, or one or more other varying properties. Heights of lines represent fitnesses of a single inheritable type. The line with the most variation in height represents changes in fitness resulting from environmental differences between small regions. Lines with less variation in height represent fitnesses relative to larger regions, treating these fitnesses roughly as averages over fitnesses in smaller regions. Originally published in Abrams (2014).

times the probability $P_A(E_j)$ of each subenvironment for a given type A. This is a mathematical expectation E of the fitnesses (see, e.g., Gillespie 2004):[7]

$$w(A) = E_j\left(w(A \mid E_j)\right) = \sum_{j}^{m} w(A \mid E_j) P_A(E_j).$$

It is represented schematically in figure 4.1. (The preceding equation is superficially similar to equation (0.1) in section 0.3.2, but equation (0.1) defines a quantity that is more closely related to $w(A|E_j)$ in the above equation.)

Levins (1968) also argued that if fitnesses are viabilities—probabilities of survival to adulthood—and each organism experiences a series of subenvironments for short periods of time with independent probabilities, then fitnesses combine multiplicatively (see also Wimsatt 1980a; Nagylaki 1992). For example, suppose that traits vary in how probable they make survival from one period of adulthood to another. Also suppose that probabilities of numbers of offspring depend only on how long an organism lives. Then traits make a difference to numbers of offspring (and hence numbers of later descendants), but only via survival. The idea is that during its life, each organism may move through or experience different environments (even if the organism's "movement" is only metaphorical, as when a predator notices it).

For example, house sparrows may eat some food in open areas that are potentially patrolled by hawks, while other food is hidden from the sky by trees. The probability of survival during periods in which a house sparrow is eating in an open area may therefore be lower than when it's eating under cover of trees. But house sparrows could have traits that influence how often they interrupt eating to scan the sky for hawks. This will affect the probability of surviving until the next time a sparrow eats.

Let house sparrows with trait A in environment E_1 (open areas, let's say) have a chance $w(A|E_1)$ of surviving. If the number of house sparrows in the population with trait A is n_A, then on average, for periods in which these sparrows are in E_1, their number will be reduced to $n_A \times w(A|E_1)$. For sparrows with A that are in environment E_2 (covered area), the number of sparrows will be reduced to $n_A \times w(A|E_2)$. So as sparrows move through different environments, their overall viabilities will have this kind of mathematical form:

$$w(A) = w(A|E_1)^{P(E_1)} \times w(A|E_2)^{P(E_2)} \times \cdots \times w(A|E_m)^{P(E_m)} = \amalg_j w(A|E_j)^{P(E_j)},$$

where the expression on the far right is just a way of summarizing the long series of multiplications to its left.[8] To understand why the viabilities $w(A|E_j)$ are raised to powers that are probabilities $P(E_j)$ of experiencing environment E_j, think of an organism's life as a series of discrete moments. At each moment, the organism is in one environment or another. On average, in any given longer period of time, it will be in environment E_j rather than another environment with probability $P(E_j)$. So the viability for moments in which the organism is in E_j is

$$w(A|E_j) \times w(A|E_j) \times \cdots \times w(A|E_j),$$

where the number of terms in this product, as a proportion of moments in all environments in the same long period, is equal to $P(E_j)$.

The additive and multiplicative cases of environmental effects on fitness that we have just examined are of course very simple, and such cases are probably rare in the real world. More realistic cases will be more complicated. Some might combine both additive and multiplicative fitness. For example, if each organism moves through a series of subenvironments during its life, but organisms experience different groups of sequential subenvironments, fitnesses relative to each member of a set of sequentially experienced subenvironments might combine multiplicatively, while the overall fitnesses relative to different sequentially experienced sets of subenvironments might combine additively. However, this is still an overly simple case in which distinct sorts of sequential variation are clearly separable from each other. The real world can be much more complicated!

Consider research by Ringsby et al. (2002), who provided indirect evidence for complex interactions between effects of weather and location in a population of house sparrows living on Norwegian islands. One might think that weather changes give us sequential environmental variation, while differences in house sparrows' locations correspond to environmental variation that tends to apply to the entire life of a house sparrow. However, much but not all weather variation is correlated between nearby islands, so the sequential variations may or may not be shared in one respect or another, and it's possible for house sparrows to migrate between islands, "switching" environments. Further, while weather patterns vary stochastically, this stochastic variation is also subject to seasonal variation, so there is a periodicity to the stochastic variation. The seasonal variation apparently affects different house sparrows' fitnesses in different ways. Whether a house sparrow is hatched early or late in the year affects what part of the annual variation it experiences during its early development, which seems to have stochastic effects on viability later. Moreover, the multiplicative scheme above only took into account viability and not also fecundity (number of offspring who survive to adulthood) or more complex variations between different subenvironments. Additional complexity would be added by mating probabilities, and perhaps other effects such as influences of parents on grandoffspring.

Bruce Glymour's work (e.g., Glymour 2011, 2014) provides additional reasons to think that fitness would have to be defined in different ways because of different kinds of environmental variation. Glymour argues that in order to predict evolution of some traits in some populations, we have to model fitness of one trait as depending on variation in environmental factors in relatively complex ways. In particular, Glymour argues that in order to explain some kinds of evolution, we have to model the causal impact of some traits on fitness as dependent on other traits or on particular environmental variations. Glymour's focus is on modeling rather than on fitness as a way of characterizing underlying properties of evolutionary processes. However, to the extent that in order to predict the effects of those processes, one must model them in ways that are not simple, the fitness relations in the process are not simple either.

Despite all of the complexity of interactions between type fitness and environmental variation, it's reasonable to think that overall fitnesses are a function of environment-relative fitnesses and probabilities of organisms experiencing environments.[9] Note, however, that these probabilities may be complicated. For example, if there is niche construction, the extent of a subenvironment could be a function of how many organisms have recently been modifying it. Thus the probability of experiencing a subenvironment could be a probability conditional on what has happened in the past with other organisms. This implies that probabilities at one time need not be probabilistically independent of probabilities at earlier times.

4.2.3 DEVELOPMENTAL PATHS

If we generalize ideas suggested by the preceding discussion, we eventually get to the following picture of the basis of causal type fitness using the idea of a population-environment system.

To a first approximation, an *organism-centered developmental path*, or a *developmental path*, is a sequence of states in the life of a *possible* organism and its environment that are consistent with the character of a given population-environment system. (The environment will of course include other members of the same population, as well as other organisms and abiotic elements with which the organism might interact.) A developmental path includes everything about a possible organism—a possible type realization—that might be relevant to the ultimate evolutionary success of the inheritable type that it realizes. A developmental path could be considered to begin at conception and end at the time of death, but paths may have to last longer when fitness is defined in terms of numbers of near-generation descendants. So, what I mean by "development" in "developmental path" includes both what is traditionally meant by that term in evolutionary biology—namely, processes by which a single cell develops into its adult state, which is often multicellular—plus all of the organism's physical changes and interactions during the course of its life and perhaps some of its descendants' lives. Note that since a developmental path includes other conspecifics, a possible sequence of states realizing a population-environment system would include overlapping developmental paths. This is unproblematic because it is the population-environment system that's primary. Developmental paths are just patterns realized in population-environment systems, and multiple patterns can be realized simultaneously.[10] The set of all organism-environment paths for a specified population-environment system and a particular biological (geno-, pheno-, etc.) type form a *developmental path space* for instances of that type consistent with the character of the population-environment system.

The point of this way of thinking about organisms within populations and environments is that it gives us a broad view of the aspects of the world to which evolutionarily relevant properties attach. A stripped-down version of this idea comes from life history theory (Stearns 1992; Caswell 2001), which models differences between organism types along dimensions of organism size, time of reproduction, clutch size, offspring size, flowering stage, and so on. William S. Cooper (2001) extends the life history idea to include arbitrary properties and environmental circumstances in organisms' lives, but he still focuses on a relatively small number of properties of lives at various times—encountering a predator, soil being hard or soft, and so on.

The house sparrow example in section 1.2 and the examples and remarks in the preceding section suggest that real-world evolution depends on interactions between an enormous number of diverse environmental and organismic properties. The large number of variables defining the dimensions of states in fully detailed developmental paths might range across sets of positions and

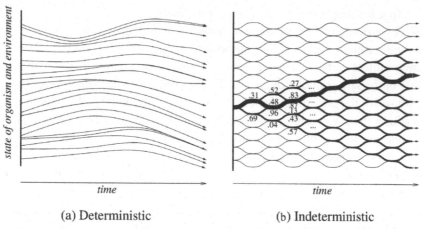

(a) Deterministic   (b) Indeterministic

FIGURE 4.2. Idealized representations of developmental path spaces assuming that biological processes
are (*a*) deterministic or (*b*) indeterministic. Only one dimension of variation in states at a time is represented, and only two-way branching is represented in *b* (where paths beginning from one initial state are
highlighted, one path is further emphasized, and probabilities of some branches are indicated).

properties of small but relatively high-level parts of organisms—both parts of
the organism in question and parts of other organisms, as well as configurations of abiotic aspects of the environment. Minutiae like the rate of flow and
cell positions in blood near a point in a vein of an organism might count as
aspects of the organism-environment state if they can make a difference to
evolution in the population. Even though it would be unrealistic to try to
capture all such variation in a model, it's worthwhile to have in mind a rich
conception of what it is that might be modeled (figure 4.2). It may sound
as if I am now taking seriously the idea of circumstance-based causal token
fitnesses that I criticized in chapter 3, but what I am discussing here are probabilities of sets of paths allowed by a given trait and population-environment
system. It's probabilities of outcomes in such sets of paths that define causal
type fitnesses. This is simply an enhancement to the picture outlined in section 4.2.2. Note that since paths form sets, this picture provides one way of
fleshing out the idea of the probability $P(\phi|\psi)$ of one possible biological event
type ϕ conditional on the realization of another event type ψ. This idea arises
in many contexts in theoretical evolutionary biology and statistical method-
ology. According the present picture, $P(\phi|\psi)$ is the (probability) measure of
paths in which ϕ occurs within those paths in which ψ occurs.

Some researchers think that biological processes are roughly deterministic (e.g., Graves, Horan, and Rosenberg 1999), as I do. Here I mean "deterministic" in a physicist's or metaphysician's sense, including the assumption that
each state of a system at a later time t_2 is determined by the state of the system

at an earlier time t_1. (In population genetics and other areas of evolutionary biology, "deterministic" has a related but different sense; see note 26, §0.3.4.) This implies that a state of the organism and its environment at one time determines the subsequent path through the developmental path space (figure 4.2*a*). Other writers think that at least some of the stochasticity in biological processes is due to fundamental indeterminism; advocates of the propensity interpretation of fitness sometimes take this position (e.g., Mills and Beatty 1979, Brandon and Caroon 1996). On this view, an initial state at conception would often determine probabilities of producing two or more subsequent states, with each subsequent state in turn determining probabilities of producing further states (figure 4.2*b*).

Either way, when I speak of the character of a population-environment system, I include not only what is possible given a particular kind of environment and population but also probabilities of developmental paths. I take population-environment systems as, by definition, specifying objective probabilities of developmental paths for particular organism types.[11] I believe that this is a reasonable assumption given the way that successful research in evolutionary biology works, as I argue in chapters 2 and 7. Chapters 8 and 9 also discuss how these probabilities might arise. In the end, my claim will be that natural selection on competing inheritable types can be understood in terms of differences in probability distributions over developmental paths for the competing types. Different inheritable traits bias the probabilities of evolutionarily relevant outcomes (e.g., number of offspring), so that for some traits, the spread of the trait is more probable—because of the influence of the traits on developmental paths.

4.3 What Variation Matters to Selection?

We now have a picture of how causal type fitnesses relative to subenvironments can combine into an overall causal type fitness. However, arguments in chapter 3 imply that fitnesses are inheritable only when particular environmental conditions can be repeatedly encountered by organisms that realize a given type. Thus, there is some sort of rough lower limit on the grain of environmental variations relative to which inheritable causal type fitnesses are environmentally determined. This is the topic of this section.

In order for organisms that are instances of competing inheritable types A_i to have fitness relative to the same subenvironment E_j, instances of each type A_i must have a significant probability of experiencing the same conditions E_j. The range of conditions corresponding to a subenvironment E_j must be broad enough to allow this. But what is a significant probability of experiencing a particular subenvironment? My previous examples and remarks imply that this depends both on the probability of occurrence of particular environmental conditions and on the probability that an organism of a given type will find itself faced with those conditions—for example, by moving into a particular part of a landscape during a certain season. But what is the cutoff value for such probabilities? Any answer, I'll argue, would depend on the relative strengths of selection and drift, and more specifically on fitness differences relative to a given subenvironment, as well as on population size. What matters is whether selection relative to a subenvironment has a reasonable chance of affecting evolution, given the other evolutionary forces acting on the population.

First, the interval of time over which evolutionary change might take place is relevant to what counts as a significant probability of recurrence. Longer intervals may produce greater probabilities of recurrence, since they allow more time for a set of environmental conditions to recur. Thus probabilities of recurrence depend on what kind of time period researchers choose to investigate (see chapters 5, 6). Second, if fitness differences relative to subenvironment E_j are small, then E_j must be encountered more often—such encounters must be more probable—for these fitness differences to have a nonnegligible impact on selection. Similarly, if fitness differences relative to E_j are large, E_j can have a probable impact on evolution even if its probability of recurrence is relatively low. Third, cutoff probabilities depend on the strength of other evolutionary forces. Consider drift, for example. Population size can determine the strength of drift.[12] If a population is small, the effect of minor differences in fitness will usually be swamped by drift. Thus, for a subenvironment to be relevant to natural selection in a small population, it may need to encompass a broader range of circumstances, ceteris paribus, to allow its probability of occurrence to be greater. These relationships could be formalized in terms of a particular model, producing a formula allowing calculation of probabilities of recurrence necessary to produce a noticeable effect of selection given values for the parameters just mentioned.[13] For a given population and environment, and a specified level of probable effect of selection, such a formula could in principle be used to estimate a *minimal environmental grain*: a specification of how fine-grained a partition of the environment can be while still making subenvironment-relative fitness differences themselves relevant to selection.[14]

I'm not sure how useful such an estimation project would be, however. The main point is that modeling practices and empirical research are consistent with a vague boundary between subenvironments that have a significant probability of recurrence and those that don't. Above this vague limit, researchers are free to choose a way of partitioning environmental conditions into subenvironments that is useful for their research goals, as suggested above and in chapters 5, 6, and 7.

4.4 Causal Type Fitness from Causal Token Fitnesses?

We saw in chapter 3 that it's a mistake to treat causal type fitness as a simple average of causal token fitnesses of those organisms actually existing in a population in a given period of time. Nevertheless, in some but not all cases, if outcomes relative to circumstances were combined in a more sophisticated way, causal token fitnesses could, theoretically, determine causal type fitnesses relative to more inclusive subenvironments that included those circumstances as possibilities.[15]

First, as remarks above suggest, we can think of a subenvironment as equivalent to a probability distribution over circumstances. The overall fitness in a subenvironment for organisms of a given type is then the probability-weighted average of the circumstance-relative fitnesses, for circumstances that can count as part of that subenvironment. For example, it might be useful to divide the house sparrow population's environment into those areas that have significant tree cover and those that are open fields. There may be a great deal of spatial and temporal variation in circumstances within either subenvironment, but some sorts of circumstances (e.g., those that include being seen by a hawk) are more likely in one subenvironment than another.

Second, a token organism can be viewed as a realization of a complex type including a whole genome and a phenotype O_i produced by the genome and whatever other factors contribute to biological development of the organism in its environmental circumstances E_j. Consider the type that consists of all properties of the organism and all properties of the environment (consistent with the population-environment system) that could make a difference to evolutionary success. Suppose that for each such complex type O_iE_j consistent with a particular inheritable trait A, there is an objective probability $P(O_iE_j|A)$ of O_iE_j occurring. Then the overall fitness of A would follow, in principle, from calculations like those suggested in sections 4.2.2 and 4.2.3. In this sense, causal token fitnesses can play a role in natural selection. Note, however, that the relevant token fitnesses are fitnesses of merely *possible* token organisms and circumstances *consistent with* the evolving genetic background of a population. Whether a possible token organism actually exists plays no role in this calculation.[16] What we are really talking about are developmental paths consistent with a given organism type and subenvironment. Each possible token organism in such and such circumstances corresponds to a developmental path (or paths, if indeterminism influences developmental paths).

Since this probabilistic averaging scheme involves probabilities of possible, not actual, token organisms, it is not undermined by my argument that arbitrary fluctuations in type fitness result from averaging token fitnesses over actual organisms (§3.3.4). The scheme can avoid some of the other problems discussed in chapter 3 as long as the types about whose evolution we want to know are ones that are relatively easy to inherit. For example, if a trait is usually the result of a few genes, it can persist over many generations, and recombination needn't quickly destroy the organismic determinants of fitness. Moreover, since it is this inheritable type that is in question, and one is averaging (probabilistically) over all of the circumstances in which instances of the type might be found, the fact that different circumstances can generate wildly different (probable) outcomes is OK; those various outcomes simply feed into the overall average of outcomes for the type. Thus probabilistic averaging thwarts the argument from recombination (§3.3.2) and the argument from circumstance variation (§3.3.3).

Let's put aside problems with hitchhiking raised in chapter 3 for the moment. Understanding causal type fitnesses as above, as probability-weighted combinations of causal token fitnesses of possible organisms in possible circumstances, we are treating causal type fitnesses as partly dependent on probabilities of realizations of types *in circumstances*, where those probabilities are determined by the population-environment system. So, although this way of understanding causal type fitness makes it depend on causal token fitnesses, causal type fitnesses are determined by much more than causal token fitnesses. Probabilities of circumstances also play a crucial role in this conception of causal type fitness. Moreover, the "possible token organisms" really are just organism-type realizations with probabilities of occurrence in various environmental circumstances *and genetic contexts*. The result is equivalent to my conception of a population-environment system. So, though we worked up to this idea starting from causal token fitnesses, what we have in the end is a concept of causal type fitness as a quantity that attaches to a type via its possible probability-weighted realizations in a population-environment systems. Hypothetical token organisms are elements of the population-environment system itself. Causal type fitnesses relative to a subenvironment are then just causal type fitnesses for circumstances (or developmental paths) that are restricted to certain locations or sets of environmental conditions.

(The argument from generalized hitchhiking in section 3.3.5 presents a different kind of challenge for understanding causal type fitness in terms of causal token fitness. It's not addressed by the preceding probabilistic averaging scheme, or by any other conception of causal type fitness that I advocate in this book. The problem is related to the one for the actual-average concept of causal type fitness, in that it has to do with the fact that linkage disequilibrium can bias a trait's probability of spreading in ways that don't depend on the trait's effects. I plan to discuss this further in future work.)

In chapter 7, I'll argue that causal token fitness plays at best a very limited role in empirical research. It may have some role in modeling, but we have to be careful about what that means.

4.5 On *the* Mathematical Definition of Causal Type Fitness

There has been a great deal of work in philosophy of biology[17] and, sometimes, in evolutionary biology that seeks a single correct mathematical definition of fitness. No doubt some readers have expected that I would propose a mathematical definition for causal type fitness, but I don't believe that it is possible or even desirable to do so. There are at least two dimensions to my skepticism. Here I merely summarize views developed further in previous and unpublished work (Abrams 2009b, Fitness MS).[18]

4.5.1 DIFFERENT QUESTIONS

First, different causal type fitness concepts can be needed for different sorts of questions about evolution by natural selection. For example, if one is interested in evolution over several generations, fitness measures must predict and explain what is likely to happen over several generations (Abrams 2009b). Such fitness measures can abstract from questions about how a population will change from one generation to the next or from one day to the next; those questions may require different fitness measures that include probabilities of very short-term changes. Questions about the probability that one trait will displace another—eventually—need not correspond to either of these.

There is a good reason for causal type fitness to be relative to questions in this way. If we think of evolution of a population-environment system probabilistically, from a given starting state of a population and environment, various possible subsequent states are possible, with different probabilities. Were we to have (hypothetically) full information about the probabilities of transitions from one population-environment state to the next, we would know much more than what any fitness measure could capture. We would know, for example, not only which inheritable trait was most likely to increase in frequency—over various time periods, and conditional on various events—but also what the chances were that another trait would succeed instead, and what the chances were that the original trait would occur in combination with some other traits (e.g., determined by other loci) in such and such frequencies at various times. We could also determine what sorts of interventions would bias probabilities of outcomes, and to what degree. Of course, the idea that we could *learn* all of this is mere fantasy. However, we should be clear about how much information about probabilities of evolutionary events is left out by any scalar fitness measure. Causal type fitness measures *summarize* probabilistic and causal facts about a population. What is useful and practical to include in the summary depends on what one is trying to understand about a population-environment system. (Potochnik [2017] and Mitchell [2009] make related points in a more general context.)

As with analogous sorts of pluralism discussed elsewhere in the book, I don't view this relativity of fitness to questions as implying that causal type fitness is unreal, merely instrumental, or *essentially* relative to our interests. Rather, different fitness concepts summarize different real causal patterns, all of which are abstractions from more complex properties of population-environment systems. Modeling and statistical inference allow scientists to infer and estimate these real properties, even though their presence or absence can't be known with certainty (chapter 7).

4.5.2 DIFFERENT CAUSAL FACTORS

Second, even relative to a specific question about evolution, what would count as a correct causal type fitness concept depends on further facts about characteristics of the population-environment system being studied, involving, for example, mating patterns, parental investment, niche construction, interactions between more or less related organisms, migration, and various sorts of environmental variation. In different population-environment systems, different properties of the system will be relevant to which type "wins" (in whatever sense of winning one has specified), but there's no reason to think that biologists have identified all of the causal factors that might turn out to be relevant in real populations. Given the complexity of population-environment systems, it seems likely that the list of factors relevant to any kind of evolutionary outcome must remain open-ended. We should therefore be skeptical of any definition of fitness that claims to have captured all of the properties of population-environment systems that are supposed to be relevant, in all contexts, to whatever questions the definition is supposed to help address.

Thus, what might be the "correct" causal type fitness concept would be relative, first, to a question asked about evolution and, second, to the sorts of complex interrelationships between organisms that will probably be relevant to that question. However, neither theoretical nor empirical research in biology needs fully accurate causal type fitness concepts in order to provide insight into evolution. Models can give insight into real-world processes that they model without the parameters of the model—in this case, causal type fitness representations—precisely describing the system being modeled. More generally, in evolutionary biology, as in much of science, fully detailed accuracy of scientific conclusions cannot and should not be expected (Wimsatt 2007; Mitchell 2009; Potochnik 2017).[19]

4.5.3 EXTENSIBLE SCIENTIFIC LANGUAGE

In the preceding paragraphs, I suggested that the list of factors that could matter to causal type fitness, relative to any particular question about evolution, has to remain open-ended. I suspect that the list of questions, too, must remain open-ended, but causal type fitness concepts may be even less constrained than the preceding remarks suggest.

Scientific language sometimes seems to develop through extensions and shifts of meaning, possibly unnoticed, as I'll argue in this section.[20] That gives scientific research valuable flexibility that it wouldn't otherwise have. Philosophical work can document such shifts and help to clarify concepts along the margins of scientific creativity, but neither philosophers nor scientists should impede progress by trying to regiment meanings where doing so is detrimental to scientific progress. I suspect this idea applies well to fitness concepts—that is, that they in have in the past exhibited, and will need in the future to exhibit, flexibility to address new research contexts.

The view that the meanings of scientific language can shift with scientists' explorations is related to Waismann's (1945) claims that scientific terms have an "open texture": that the meaning of a single term and the evidence for its applicability can't be specified once and for all but must be extended for new contexts.[21] That is, *the* or *a* meaning of a word isn't fixed but can develop over time in response to our interactions with the world. I think that something like that is correct—but what I am claiming here is slightly different: it's that a single word can develop new and distinct but related meanings as scientists come to use it in new ways. I'll start with an argument by Evelyn Fox Keller against the view that metaphors and other figurative language should be considered merely temporary heuristics in science, to be dispensed with as soon as possible:

> Yet, as historians and philosophers have increasingly come to appreciate, close observation of scientists at work, either in the present or in the past, reveals that they simply cannot function under such a harsh mandate. The difficulty

is obvious: scientific research is typically directed at the elucidation of entities and processes about which no clear understanding exists, and to proceed,
scientists must find ways of talking about what they do not know—about that
which they as yet have only glimpses, speculations. To make sense of their
day-to-day efforts, they need to invent words, expressions, forms of speech
that can indicate or point to phenomena for which they have no literal de-
scriptors. As Mary Hesse [(1987)] reminds us, "The world does not come
naturally parceled up into sets of identical instances for our inspection and
description." [311] Making sense of what is not yet known is thus necessarily
an ongoing and provisional activity, a groping in the dark; and for this, the im-
precision and flexibility of figurative language is indispensable. (E. F. Keller
2002, 118)

Keller's point can be generalized beyond obviously figurative language: Scientists can also use existing, nonfigurative scientific terms in new ways that
enlist the terms in related but novel uses.[22] On this view, scientific language
is a part of natural language, and many natural-language terms have ambiguous and vague definitions. Terms apparently evolve over time through use
by people in their daily lives; dictionary editors struggle to keep up.[23] Why
wouldn't science partake of the same linguistic freedom if doing so allowed
application of conceptual frameworks and models to new contexts? (When
there are problems with this strategy, scientists may resort to regimentation
of language, very occasionally with philosophers' help.)

Here are a few simple illustrations of linguistic shifts from biology. In each
case, there has been a more or less subtle shift from one meaning or use of a
term to conceptually similar uses that accord better with particular researchers' needs. (I'm not implying that scientists are necessarily aware that a new
meaning has been created or that a precise sense has been given to a term
through its new use.)

1. Lynch and Walsh (2018, 474–75) remarked that, originally, "*cis*-acting"
genetic elements were ones that modified the phenotypic expression of a gene
sitting on the same piece of DNA, while "*trans*-acting" genetic elements could
affect the expression of genes anywhere in the genome. However, researchers
investigating these modifiers "co-opted" (Lynch and Walsh 2018, 475) these
two terms to refer to genetic elements that acted on specific nearby or specific
distant genes, respectively. That is, the basis for the distinction was no longer
whether or not the influence was restricted to the same chromosome.

2. In 2005, Weinreich, Watson, and Chao defined "sign epistasis" to refer
to a genetic variant (e.g., an allele at a single locus) whose effect on fitness
depends on genetic patterns at other loci in such a way that it is "beneficial on
some genetic backgrounds and deleterious on others" (Weinreich, Watson,

and Chao 2005, 1166). That is, originally, "sign epistasis" was the property of a genetic pattern that would cause either increased or decreased *fitness*, depending on which other genetic patterns an individual had. Subsequently, some authors used "sign epistasis" to refer simply to a phenotypic influence that can be reversed by differences in genetic background, regardless of its fitness consequences. For example, Natarajan et al. (2013) wrote, about the "affinity" of hemoglobin for oxygen, that

> the H-type β-globin allele conferred an increased affinity on the αLL background and a decreased affinity on the αHH background. These are examples of sign epistasis [(Weinreich, Watson, and Chao 2005)], where the sign of the phenotypic effect of an allele is conditional on genetic background. (Natarajan et al. 2013, 1325)

The idea is that a particular allele can either cause more interaction between hemoglobin and oxygen than alternative alleles cause, or less interaction than those alleles cause, but this difference in direction depends on genetic background—on alleles at another locus. We can see that Natarajan et al. don't view this kind of sign epistasis as intrinsically involving fitness differences, since they explicitly mention its effects on fitness, represented by selection coefficients (e.g., Gillespie 2004):

> The pervasiveness of sign epistasis for Hb-O_2 [hemoglobin-oxygen] affinity suggests that the selection coefficient for a given allele will often be highly dependent on the allelic composition of the local population. (Natarajan et al. 2013, 1326)

Notice that in the earlier quotation, Natarajan et al. cited Weinreich, Watson, and Chao (2005) as a source of the term "sign epistasis," despite the latter's restriction of the term to fitness reversals. This sort of shift of meaning from cited source to current application is not routinely treated as problematic, as far as I can see. The next example illustrates the same pattern.

3. Neel (1962) is widely cited as the originator of the "thrifty gene hypothesis," usually treated as a proposed explanation for why some but not all people have genes that make obesity more likely in modern industrialized societies (e.g., Speakman 2013). Neel suggested that there might have been prior selection for "thrifty genotypes" that favored physiological mechanisms that conserve energy "when food intake was irregular and obesity rare" (Neel 1962, 358). Though Neel mentions obesity, his hypothesis was proposed as an explanation of genetic variation of influences on diabetes, not obesity. However, because similar "thrifty" genotypes could cause obesity as well as diabetes, papers that focus on evolution and obesity routinely use "thrifty gene

hypothesis" to refer exclusively to a proposed explanation for obesity-related genetic variation, often citing Neel as the source of the idea.[24]

In none of these three cases is there any reason to think that the shift in terminology has led to confusion among scientists. Nersessian (2008) also described cases in which scientific concepts were developed gradually during experimental and theoretical research. Wilson's (2006) discussion of shifts of usage of terms in physics between very closely related contexts is also relevant.

Of course, not all cases of pluralistic language in science must be as flexible as what I have in mind. Some species pluralists such as Ereshefsky (1992) argue that there are a few relatively well-defined alternative species concepts. Similarly, Griffiths and Stotz's (e.g., 2013; 2014) work on gene concepts seems to imply that there are only a few widely used gene concepts, while Waters (2017, 2019) argued that some gene concepts are parameterized by clearly specifiable variables to allow applications in different contexts. None of these cases seem to involve the same sort of creative flexibility that my discussion describes. In chapter 5, I'll provide further illustrations of creative shifts of meaning, and I will use them as the basis of a view about roles of populations in evolutionary biology.

I argued in preceding sections that it's not possible or useful to try to work out *the* correct causal type fitness concept. However, if scientific language often changes through small shifts of meaning to exploit new contexts, there's even less reason to think that it's possible to work out the one true concept of causal type fitness. The multiplicity of existing uses of "fitness" suggest that it might be exactly the kind of term that is deployed with subtly different meanings for different contexts. This doesn't mean that philosophers can't work to clarify or reconstruct specific ideas about fitness, of course. It just means that in doing so, they should probably pay attention to and often defer to the roles of the concepts in *particular* research contexts or in historical development. My five-way classification of fitness concepts is an attempt to provide a lightweight, clarifying structure for the panoply of fitness concepts. It may also provide guidance for more detailed research on the conceptual development of fitness concepts.

4.6 Conclusion

In chapter 3, I argued that causal token fitnesses of actual organisms are not the fundamental basis of natural selection and that views that treat causal fitnesses of actual token organisms as fundamental cannot make sense of selection's basic character. In this chapter, I presented a view of causal type fitness as involving a possibly complex function of probabilities in subenvironments relevant to numbers of descendants, along with probabilities of organisms with inheritable types experiencing each subenvironment (§4.2, §4.3). This scheme allows an environment to be partitioned into subenvironments in various ways, and it allows subenvironment-relative fitnesses to be derived from narrower component subenvironments. However, I argued that those subenvironments that are unlikely to be experienced repeatedly don't determine a sense of fitness that allows comparisons between competing types (§4.3). This constraint depends on various factors, including researchers' interests. I explained that it may be possible to define type fitness as a function of token fitnesses in some situations, but only by using objective probabilities over merely possible token organisms in possible circumstances (§4.4). Such probabilities ultimately depend on the character of the population-environment system in which a population is evolving, so the ostensibly derived type fitnesses do as well. Finally, I gave several reasons why it's a mistake to think that identifying a unique, precise definition of causal type fitness is a worthwhile project (§4.5). The next few chapters clarify ideas about population-environment systems and natural selection, and further clarify the roles of fitness concepts in empirical research.

PART II

Reconstructing Evolution and Chance

Part I laid the groundwork for the views that I develop in the rest of the book. Chapters 1 and 2 introduced the idea that evolution takes place in *population-environment systems* exhibiting *lumpy complexity*, where each population-environment system is a *chance setup* realizing *causal probabilities* of outcomes within that population-environment system. Those outcomes include changes in frequencies of traits realized by organisms in a population-environment system. It's probabilities of such outcomes, I argued, that are the foundation of natural selection (and drift). Chapter 2 also explained that the argument I gave that population-environment systems realize causal probability is an instance of an *argument from empirical practice*. This kind of argument plays a central role in some of the remaining chapters, which will include a second argument for causal probabilities in population-environment systems and an extended discussion of their nature. Chapters 3 and 4 were partially but not entirely destructive, in the sense that I argued that views about biological fitness common in philosophy of biology either don't fit the needs of evolutionary biology or don't make sense of what I take to be general truths about evolutionary processes. I introduced a four-way distinction between kinds of fitness concepts—*causal token-organism fitness, measurable token-organism fitness, causal organism-type fitness,* and *statistical organism-type fitness*—as well as a fifth category of *purely mathematical fitness*. Philosophers of biology have often treated causal token-organism fitness as fundamental, but in chapter 3, I argued that it has little relevance to evolutionary biology. Chapter 4 began to develop an alternative view by discussing different ways that environmental variation is relevant to natural selection, suggesting that causal organism-type fitness is the fundamental basis of natural selection, and explaining why this kind of fitness need not have a single definition.

Part II also includes criticisms of previous views, but its focus is on building upon the ideas introduced in part I to develop a positive view about what natural selection and evolution are. In chapter 5, I argue for a kind of population pluralism, according to which biologists do and should have great flexibility in the way that they define biological populations. Some philosophers think that population pluralism undermines the idea that natural selection is a cause of evolution. In chapter 6, I argue that it doesn't undermine this idea, because of the way that causal patterns in complex systems depend on aspects of those systems picked out for study. Chapter 7 returns to discussion of fitness concepts, and I argue that biologists use measurable token-organism fitness and statistical organism-type fitness to make statistical inferences about causal organism-type fitness, and that it is the latter that is fundamental to evolution. Causal token-organism fitness—the kind that many philosophers have thought fundamental—plays a role in evolutionary biology only in very special senses. Finally, in chapters 8 and 9, I investigate the nature of the causal probabilities realized in population-environment systems. I argue that some previous proposals about the character of probabilities involved in evolutionary processes are unlikely to be correct. I explore the idea that these probabilities are *measure-map complex causal structure* probabilities, a kind of probability previously proposed by several authors, including me.

Populations in Biological Practice:
Pragmatic Yet Real

5.1 Introduction

In chapter 1, I introduced the idea of a population-environment system and suggested that it is within such entities that evolution takes place. However, I didn't specify how to identify population-environment systems. Clearly, specifying a population will be a big part of this. In this chapter, I'll look at ways of defining "population" that philosophers of biology have proposed, and I will argue that these strategies sometimes fit poorly with established, fruitful practices in evolutionary biology. My view is that "population" corresponds to a family of shifting, related, and sometimes vague concepts that are nevertheless able to delineate groups within which natural selection and other aspects of evolution take place. This fits well with the view of scientific language that I sketched at the end of chapter 4. Beginning in this chapter and continuing in chapters 6 and 7, I explain how an open-ended, pluralistic approach to populations can nevertheless support an understanding of evolutionary forces as causes of evolution.

After some initial discussion of population pluralism in section 5.2, in section 5.3 I present and criticize the view that we should try to develop one or a few simple definitions of population concepts. I focus especially on Roberta Millstein's work, as she has published more on this topic than anyone else. Among other things, I argue in these sections that biologists need only a minimal notion of population in order to ask research questions that would be precluded by definitions like Millstein's. I use examples of research on long-term human evolution to argue that biologists deploy population notions in very flexible ways, and that this is a fruitful practice. Then, in section 5.4, I discuss an example of an even more flexible use of ideas about populations in a study of recent short-term human evolution. Finally, in section 5.5 I suggest that rather than trying to define a fixed or parameterized set of more or less precise population concepts, it might be better for philosophers of biology to investigate ways in which ideas about populations reflect particular research contexts or their historical relationships.

5.2 Population Pluralisms

The most basic, seemingly trivial requirement for something to be considered a population of organisms is that it be the sort of thing in which evolution could take place. We can't require for any population that evolution must take place in it or that any particular kind of evolutionary process—natural selection, drift, mutation—must take place in order for something to count as a population. After all, a population has to be something about which it makes sense to ask, and to investigate, whether evolution takes place and why. Still, at minimum, a population must contain some organisms that might plausibly be able to reproduce and pass on their traits. What else, if anything, must be true of a collection of organisms in order for it to count as a population?[1]

Recent philosophical discussion about how to think about biological populations has sometimes been framed as a debate between different sorts of "population pluralism"—claims that evolutionary biology allows many concepts of population—or between pluralism and "population monism," according to which a single population concept is all that's needed, or the view that there are only a small number of legitimate population concepts. Causalists such as Millstein (2009, 2010a, 2010b, 2014, 2015), Otsuka et al. (2011), and Gildenhuys (2014) agree with statisticalists such as Denis M. Walsh (2007, 2013) and the population pluralist Stegenga (2010, 2016) that if there is a great deal of flexibility about what can count as a population, there might be no fact of the matter about whether natural selection or random drift cause evolutionary change. For example, Walsh assumes that populations can be defined in many ways, and he uses this thesis to argue that natural selection and random drift are not causes of evolution (see chapter 6). By denying Walsh's assumption that populations can be defined quite flexibly, Millstein, Gildenhuys, and Otsuka et al. seek to undermine such arguments.

Although my view is causalist, I'll argue for a pluralist view that is related to Gannett's claim that what counts as a population depends on research contexts:

> Populations are pragmatically and variably constituted in different sorts of investigations of species genome diversity. Genes become bounded in space and time in ways that fulfill aims, interests, and values associated with particular explanatory contexts. Population boundaries are not fixed but vary from one context of inquiry to another. (Gannett 2003, 990)

I think that something like this view best describes a great deal of very fruitful practice in evolutionary biology. By contrast, Millstein argues that evolutionary biology would benefit from the addition of a central, clearly defined population concept; she cites review articles in which biologists describe conflicts due to misunderstandings about what count as populations. Note that while I will begin by discussing population concepts, I think that putting the question in terms of distinct population *concepts*, as many authors do, has the potential to lead to misunderstandings about roles of populations in practice in evolutionary biology—as I'll explain later.

In an earlier publication, I suggested that "there is no one concept of population appropriate to all contexts in evolutionary theory" (Abrams 2009c, 26), but I provided a loose necessary condition for populations:

> If there is at least a small but significant probability that there will be interbreeding and genetic recombination between an organism *o* or its descendants and a member of the population at some point in the interval of time under consideration, then *o* is also a member of the population. (Abrams 2009c, 26)

I now think this requirement—implicit in views like Millstein's—is too restrictive.

5.3 Against the Definitional Strategy

5.3.1 MILLSTEIN'S DEFINITION AND OTHERS

In a series of papers, Millstein (2009, 2010a, 2010b, 2014, 2015) has refined and explored implications of what she calls the *causal interactionist population concept* (CIPC) definition of "population."

- Populations (in ecological and evolutionary contexts) consist of at least two conspecific organisms that, over the course of a generation, are actually engaged in survival or reproductive interactions, or both.
- The boundaries of the population are the largest grouping for which the rates of interaction are much higher within the grouping than outside (Simon 2002). (Millstein 2010a, 67)

Thus, the boundaries of a population occur where there is a drop-off in the rates of certain actual interactions. Here is a succinct summary, from a later paper, of Millstein's sense of "survival or reproductive interactions":

> Reproductive interactions include both unsuccessful and successful matings (interbreeding), as well as offspring rearing. Survival interactions are almost as broad as Darwin's "struggle for existence," including competition for limited resources as well as cooperation. Social interactions are not a separate

category, but may fall into either or both of the other two categories; social interactions that do not affect survival or reproduction are not relevant evolutionarily or ecologically. (Millstein 2015, 6)

Millstein's idea of drawing boundaries where interaction rates drop off is derived from Herbert Simon's (1996, 2002) notion of nearly complete decomposability or near decomposability (Millstein 2009, 2010a). Millstein expresses openness to some sorts of pluralism but has repeatedly argued for the relevance to evolutionary biology of the preceding concept. I do think that Millstein's population concept (like others I discuss) captures some ideas that are important in evolutionary biology, but I'll argue that the CIPC is inconsistent with many roles that populations play in evolutionary biology. (Matthewson's [2015] definition of what he calls "paradigm Darwinian populations" is closely related to Millstein's CIPC, differing mainly in details, and most of my criticisms of Millstein's view can easily be adapted to Matthewson's.[2])

Millstein is right to emphasize the importance of causal interactions affecting survival and reproduction to the role that populations play in evolutionary thinking. These are interactions that matter to the evolution of a population. It's appropriate that Millstein's definition is intentionally vague about how much of a drop-off of interactions might define a population, as this allows the concept some flexibility for different research contexts. Millstein is also intentionally vague about exactly what might count as survival or reproductive interactions. This also allows the idea to have the flexibility to apply to different cases that might come up in actual research.[3]

Barker and Velasco (2013) argued that in practice, biologists would have to make choices not only about what counts as a sufficient drop-off in strengths of interactions but also about how to weight factors that contribute to survival or reproductive interactions. For example, survival might depend both on competition between organisms for a food resource and on cooperation between organisms in defending against a predator. Both of these count as survival interactions. Suppose that conspecific organisms A_i engage extensively in *both* sorts of interactions with each other, and that conspecific organisms B_j also engage extensively in *both* sorts of interactions with each other. Suppose that A_is and B_js also engage in extensive *competitive* interactions with each other, but rarely engage in *cooperative* interactions. On Barker and Velasco's view, a researcher focusing on the role of cooperative interactions in evolution should consider the A_is and B_js to be members of different populations, because there is a drop-off in those interactions that matter to the research. A researcher primarily interested in competition for food, on the other hand, might be better off considering all of the organisms to be members of a single population. Thus, Barker and Velasco suggested that what counts as a population according to a definition like the CIPC should depend on decisions that take into account research goals and other contextual factors. This accords with Gannett's point that populations are and should be defined in ways that fulfill "aims, interests, and values associated with particular explanatory contexts" (Gannett 2003, 990). Note, however, that in the preceding example, researchers interested in the influence of cooperative interactions over a sufficiently long period of time may also need to take into account the interpopulation competitive interactions, since even if these interactions are rare, they could influence the evolution within the two populations over the long haul. This agrees with Gannett's point as well. (It should become clear below that what I mean by "take into account" doesn't necessarily require that researchers must treat all of the competitively interacting organisms as members of the same population.)

There are certainly research contexts in which "population" is used in ways that fit Millstein's CIPC (Millstein 2010a) or Barker and Velasco's view. I can easily imagine that some biologists would find it useful to conceptualize populations in this way. The multidimensional vagueness that is inherent in Millstein's CIPC and that has been extended by Barker and Velasco gives a concept like the CIPC flexibility that can allow it to provide a useful way of thinking about populations in various contexts.

5.3.2 ADDING PROBABILITY

I'll explain in later sections that I don't think that the CIPC is on the right track as a general characterization of populations studied in evolutionary biology. It will nevertheless be instructive to see how the CIPC can be incrementally improved by incorporating probability into its definition. Millstein's requirement that organisms *actually* interact seems to fit poorly with factors that causalists (and, I believe, evolutionary biologists) consider important to understanding evolution. Seeing how to avoid this problem by adding probability will give us another entry point to the population-environment conception of evolutionary processes that I'll describe later. (A similar improvement could be applied to Matthewson's definition.)

Suppose that there are significant (causal) probabilities of organisms A_i and B_j interacting with each other but much lower probabilities of them interacting with organisms C_k. Suppose that, nevertheless, by chance the A_is barely interact with the B_js at all, but the B_js happen to interact extensively with the C_ks. Then despite the original probabilities, the B_js and C_ks have turned into a population according to the CIPC, excluding the A_js. Now, either on the propensity interpretation of fitness view that causal token fitnesses are fundamental or on my view that causal type fitnesses are fundamental, causal fitnesses depend on chances of outcomes such as reproductive events. However, probabilities of these outcomes will often depend on probabilities of interactions with conspecifics. So by treating population divisions as dependent on actual rather than probable interactions, the CIPC implies that causal fitnesses can depend more on probabilities of interactions across population boundaries than on those within population boundaries—just by chance. That is, either on the propensity interpretation of fitness view or on my view, fitnesses are defined by probabilities of interactions. But after the fact it may turn out that the probable interactions that contributed most to fitness were probable interpopulation interactions.

Moreover, Millstein holds that what counts as an environment depends on what counts as a population: "the boundaries of the environment would be delineated by the fullest extent of the spatial location of the population" (2014, 749) (see chapter 6). When there happen to be many more actual but improbable interactions than there are actual probable interactions, this redefines what counts as an environment, on Millstein's view. In such situations, the CIPC will paradoxically imply that fitness (in a causal, probabilistic sense) depends less on conditions within a population's environment than on those external to it: fitness depends on probabilities, but the environment need not, it seems. Thus fitness wouldn't depend essentially on either the environment or on the population! Note, too, that if chancy interactions between conspecifics are part of drift (Beatty 1984; Millstein 2002; Gildenhuys 2014), the CIPC implies that the effects of drift can change population boundaries. Researchers might then have to know whether drift of certain kinds occurred in order to know whether there was a population—about which one could then ask whether drift occurred (see §5.3.3 for more on this point).

If we wanted to pursue a strategy suggested by Millstein's approach, we could try to define a modified CIPC, such as this *probabilistic causal interactionist population concept* (PCIPC):

- Populations consist of conspecific organisms such that—over the course of m generations or an interval of time I—either the organisms or their descendants would (causally) probably be engaged in survival or reproductive interactions, or both.
- A population is the largest grouping for which the probabilities of interaction are much higher within the grouping than outside.

This definition eliminates the problem that randomness in what interactions actually take place can affect what counts as a population; even if members of two sets of organisms in fact don't interact with each other, if it was probable that they would, then they would be part of the same population. The PCIPC also restores the idea that causal senses of fitness depend primarily on intrapopulation interactions and intraenvironmental factors, since what actually happens by chance cannot create new population and environmental divisions that make fitnesses depend primarily on cross-boundary factors. To see the relationship between the CIPC and the PCIPC, notice that actual interactions, which are the basis of the CIPC, are usually those that are most probable according to the PCIPC. The actual interactions can therefore usually be taken as evidence for what is probable. However, when many improbable interactions occur, the CIPC may divide populations differently than the PCIPC.

In the definition of the PCIPC, I relativized the probabilities to generations or time because these can make a big difference to the probabilities of interactions (see chapter 6 and Abrams 2009b). Organisms that have a low probability of interacting now might have a high probability of having descendants that interact. Consider an example I discuss below and in chapter 7 in more detail: *E. coli* bacteria that don't compete for food because there is a great deal of food available may have descendants that compete for food when nutrients in the culture in which they live become depleted (see §7.3.2). One might think that a solution would be to require that only short time intervals or a few generations be allowed to matter for the definition of a population. However, as the *E. coli* example shows, what counts as a population defined in terms of a short time constraint could be a poor basis for thinking about future evolution. So a better view is simply to relativize probabilities of interactions to a period over which evolution might take place (Abrams 2009b). This means that different sets of objective probabilities could be relevant to different research questions (see below and chapter 6).

The PCIPC, like the CIPC, is illuminating because it captures pragmatic factors that *can be* relevant to defining a group of organisms as a population. There may well be particular research contexts in which it would be useful for a group of biologists to adopt the CIPC or the PCIPC. The PCIPC also helps to emphasize what role probability should have in our understanding of population-environment systems. However, the rest of this chapter constitutes a series of arguments that the definitional strategy illustrated by the CIPC, the PCIPC, and proposals of some of the other authors mentioned here is not a good general strategy for understanding roles of populations in evolutionary biology.

5.3.3 POPULATIONS AS PRIOR TO ANSWERS

Let's look more closely at experimental evolution with *E. coli* bacteria—in particular, at some of the celebrated research by Lenski and his colleagues. In Lenski et al. (1991), different *E. coli* "populations" (their term) were grown, each from a single bacterium. The populations were physically isolated from each other. There were no reproductive interactions between bacteria within a single population,[4] but there was apparently competition for food: bacteria were transferred to a new culture every day, because otherwise their food would have become depleted. It appears that at least before the transfer, there may have been some competition for food, but a small change in the procedure could have shifted the bacteria to new cultures more rapidly, so that there was almost never competition for food. In this case, researchers would apparently be studying evolution, but according to the CIPC and the PCIPC, this would be evolution in a nonpopulation, since no bacterium would interact with any other.[5]

A problem with definitions like the CIPC and the PCIPC is that they don't allow researchers to define a population *and then* investigate whether there are interactions or particular probabilities of interactions between the members of the population (see Stephens [2010] for related discussion). What if the scientists performed an *E. coli* experiment without food competition *in order to find out* whether there would be interactions of some other kinds between the bacteria? Millstein's definition implies that researchers should withhold the word "population" until they find the answer.[6] However, it has to be possible to define populations—or things that are treated like what researchers call "populations"—prior to knowing the answer to questions about interactions between members of the population.

More subtly, the CIPC and the PCIPC don't allow researchers to define distinguishable groups of organisms as populations and then investigate how much interaction there is (or how probable it is) *between* these "populations." Consider these remarks concerning models involving migration as well as selection:

> In general, theoretical models point toward a critical migration threshold above which selection will be overwhelmed by migration and no local adaptation is possible (i.e., allelic "swamping" happens; Lenormand 2002). Alternatively, below this threshold, selection may overcome the effects of migration and the beneficial polymorphisms are maintained in a set of *populations*, resulting in local adaptation (Antonovics and Bradshaw 1970; Hendry et al. 2002; Sambatti and Rice 2006; Gonzalo-Turpin and Hazard 2009; Comeault

et al. 2015; Monnahan et al. 2015). (Hämälä, Mattila, and Savolainen 2018; emphasis added)

The thing to notice about the quotation is that it describes models that incorporate migration and selection but differ in how strong those two processes, as modeled, are in relation to each other. Small amounts of migration would mean that there is little interaction between organisms in different populations (Hämälä, Mattila, and Savolainen's term), and large amounts of migration mean would mean a great deal of interaction between organisms from different populations. Thus, according to the CIPC or the PCIPC, whether a model counted as a single-population or multiple-population model would depend on the values of parameters that (also) determine the strengths of selection and migration.

The paper from which the above quotation comes describes an empirical study, however. Hämälä, Mattila, and Savolainen (2018) described their study of the evolution of eight collections of *Arabidopsis lyrata* plants that were defined by locations in two regions of Norway. (There were also two control populations, from Germany and North Carolina.) The researchers then estimated amounts of migration between these populations (as the paper calls them), along with drift and selection within each population. An important question was whether the effects of migration and drift would swamp the effects of natural selection, and it turned out that selection still had a measurable effect, despite the amount of migration and drift. This is an example in which researchers are clearly studying evolution in populations defined in ways that were convenient. Whether these particular populations satisfied CIPC (or PCIPC) requirements seems like an additional, superfluous question. Requiring that a group of organisms engage in (or have significant probabilities of engaging in) particular sorts of interactions seems contrary to the needs of such empirical research, and to the way that it is in fact carried out.[7]

5.3.4 STUDIES OF LONGER-TERM HUMAN EVOLUTION

Many uses of "population" in empirical research make populations dependent on what data sets are available in practice, and on mathematical constraints on statistical inference given characteristics of the data. This illustrates another dimension of the flexible and pragmatic uses of "population." In many of these cases, populations are defined as they are partly because researchers do have reasons to think that there are more causal interactions

and probable causal interactions within than between (what they call) populations. Nevertheless, both the CIPC and the PCIPC would be too inflexible
for such research contexts. I'll illustrate my points using an active area of
research on human evolution using large genomic data sets. The examples
I'll discuss use two data sets: HapMap[8] and the Human Genome Diversity
Cell Line Panel (HGDP-CEPH, or HGDP)[9] data sets, each of which contain
genetic data from various parts of the world. Both data sets have been widely
used for investigating human evolution, although HapMap has been retired.[10]
In the area of research on which I'll focus, various statistical tests on genetic
data are performed to determine whether certain parts of the human genome
have been under selection during the last tens of thousands of years or during
earlier periods. Similar research has been done on many other species (Haasl
and Payseur 2016).

To take one illustration, Moreno-Estrada et al. (2008) started with DNA
data from 1064 people "representing 51 populations" (144) from HGDP-CEPH
data.[11] Then, "in order to maximize sample sizes, population samples were
regrouped into 39 populations based on geographic and ethnic criteria" (145)
(specified by Gardner et al. 2005). A subsequent paper by some of the same
authors used the same "worldwide sample of 39 human populations" (Moreno-
Estrada et al. 2009, 2286). Like many researchers who use HGDP, Moreno-
Estrada et al. (2009) also collected these small populations into larger groups
corresponding roughly to seven continents, referring to these larger collections of individuals as "groups" (2288, 2292), "continental regions" (2288–2290),
"continental levels" (2289), or "geographic regions" (2286, 2289–2292). Further,
Moreno-Estrada et al. (2009) compare some of their HGDP results for combinations of subsets of the thirty-nine populations with large "populations" from
the HapMap data set:

> Moreover, two of them (i.e., rs4245224 and rs628831) reproduced exactly the
> same DAFs [derived allele frequencies] of rs2229177 from the four HapMap
> *populations* in our equivalent samples from the HGDP-CEPH panel (i.e., Yo-
> ruba, French, Han, and Japanese). (Moreno-Estrada et al. 2009, 2293; empha-
> sis added)

I include the quotation not for the details in the first half of the sentence,
which describe evidence for natural selection at particular loci.[12] The thing to
notice is that when discussing the HapMap data, Moreno-Estrada et al. use
"population" to refer to a larger group of humans than when they only discuss
HGDP.

Thus, we have the same authors using "population" for each of fifty-one
groups of humans represented as distinct, genetically related groups in the

HGDP data, as well as for thirty-nine groups that unite some of those fifty-one populations, and also for even larger groups of humans represented by a different set of samples in the HapMap data. Moreover, the reason that the fifty-one populations were grouped into thirty-nine populations was simply "to maximize sample sizes"—that is, to facilitate statistical tests, rather than because of something about the populations that were grouped together. One can find similar patterns for the use of "population" in other papers using HGDP, HapMap, and other genetic data sets. With HGDP, one often sees a distinction between small "populations" and larger "groups" of data from them, but that wording seems to be used for the sake of clarity in an article that discusses both. In another context, authors might call the continental groups "populations," as in the quotation above.[13] Reading papers like Moreno-Estrada et al.'s, one gets no impression that regional groups and smaller groups are treated in a fundamentally different manner, and no special difference marks groups that happen to be designated as "populations." The questions and tools that are used to study both the "populations" and the continental "groups" are often exactly the same: Was there selection on this part of the genome, or is the pattern we see the result only of drift, recombination, changes in population size, or other factors? What do various statistical tests indicate?

For example, Moreno-Estrada et al. (2009) used a test for selection known as XP-Rsb. They compared both XP-Rsb values at "regional levels" and also in smaller, component "populations":

> Previously observed signals at the *regional level* could be attributed to specific *populations*. . . . As for the VPS37C gene (fig. 1e), we found that significant XP-Rsb values were obtained in this gene region for three of six East Asian *populations*, namely, Northeast China, Han, and Japanese, with Han Chinese accounting for most of the significant values. Not surprisingly, the signal is less significant than the one observed at the *regional level*, because the latter is decomposed into different individual *population* signals. (Moreno-Estrada et al. 2009, 2292; emphasis added)

This text reflects the fact that the authors found that XP-Rsb values at many single-nucleotide polymorphisms (SNPs) near the gene VPS37C provide statistically significant evidence of natural selection (Moreno-Estrada et al. 2009, fig. 1d, 2288) in the continental East Asian (EASIA) "regional" group, and they then found that statistically significant XP-Rsb values appeared in the Northeast China, Han Chinese, and Japanese component "populations" but not in the other three component populations of the East Asian regional "level" (Moreno-Estrada et al. 2009, fig. 1e, 2288).[14] Notice that the same sort of evidence for natural selection is used for groups of various sizes, whether they are called "populations" or not.

Thus in at least one kind of empirical research in evolutionary biology, evidence is found for evolution in groups, in which the same group may be called a "population" or something else in different studies, but where the boundaries between these groups are is not clear.[15] I see no reason to assume that the different Chinese groups, for example, would satisfy the CIPC. Such practices have been exhibited in many papers, apparently without causing conflict or confusion among the scientists concerned with this research. So, in this area, "population" is used in very flexible ways. Such "populations" do depend to some extent on probable past interactions between individuals; no one would group the Yoruba and Japanese data sets together as a population. However, the degree of past interaction doesn't need to be known in order to treat a group of people as a "population."

Thus, scientists claim to find evidence of natural selection in populations that don't satisfy definitions of "population" discussed in sections 5.3.1 and 5.3.2; what these scientists call "populations" are often mere pseudopopulations according to those definitions. If one wanted to maintain that one of the definitions was correct, perhaps one should argue that researchers do not in fact find evidence of natural selection in such pseudopopulations when they claim they do. Or one could claim that the "real" population is larger than some of the groups treated as populations by researchers. In that case, it's not clear that it makes sense to allow that there can be evidence of natural selection in some parts of the larger population but not others, as the quotation above from Moreno-Estrada et al. (2009) implies. Or perhaps one could insist that the smaller "populations" are merely subpopulations, not *real* populations, with selection in some subpopulations but not others. However, since natural selection in subpopulations can be investigated in exactly the same ways as in the larger population, such a terminological distinction seems superfluous. Given the amount of research that uses "population" in ways that conflict with the definitions discussed above—research that appears to be quite fruitful—those who want to defend a definitional strategy have a severe burden.

(Spencer [2016] uses fuzzy set theory to provide a sophisticated ontology for what he calls "K populations." Many of the populations discussed in this section count as K populations, and Spencer's account derives from detailed analysis of work like Moreno-Estrada et al's. Spencer's account could be used to fill in details for some points I make, but doing so would require a digression that's not warranted for my purposes. Moreover, while it can be illuminating to see how a particular set of real-world population-classification practices can be given a mathematically precise ontology, Spencer does not claim, nor should we, that his system characterizes all uses of "population." According to the view that I advocate in this chapter, any overly precise constraints on population concepts will not do justice to their productive uses by scientists, and if adopted by scientists, could impede scientific progress.)

5.3.5 WAYS OF MODELING CAUSAL CONTRIBUTIONS

What's called a population will often reflect causal interactions that matter—or rather, could, probabilistically, matter—to actual evolution in it. That is the point of characterizing a group of individuals as a population. It is the idea in the CIPC that I think is correct. What this means is that in studying a population, one would attempt to find evidence of causal interactions that might matter to evolution and/or find that some interactions are *sufficiently negligible*—which is not to say negligible in an absolute sense—that one can ignore them and get evidence of other (approximately characterized, perhaps in a summary way) causal interactions and their causal probabilities that help explain what happens (did happen, will happen) in a population. For this purpose, it is convenient—even, perhaps, ideal—that the population picked out by biologists' delineation of it turns out in the end to be one that satisfies the PCIPC (and ideally, the CIPC as well). For a PCIPC population is one in which causal probabilities of interactions across its boundaries are small, and hence can be profitably ignored.

But satisfying the PCIPC is not necessary. For example, if conspecifics probably routinely move in and out of a population, engaging in survival and reproductive interactions with those in the populations, that population is not a PCIPC population. However, those interactions can then be modeled as due to "migration," and the population can be treated as a system in which the "environment" consists in part of effects of those who move in and out of the environment, and in which genetic influences are modified by those migrations. In this population, or population-environment system, natural selection can take place, for example, and this fact can be modeled and investigated.[16] (I give an example of an extreme case of this below.)

As I will argue later, this is not to treat "population" as a mere epistemological construct—as an idea that is relevant to modeling and empirical research but not to the nature of processes in the world. In fact, my claim will be that when biologists model things they call populations, and find evidence of natural selection in them, they are—imperfectly, approximately, with defeasible evidence—discovering causal patterns in a real system that is something like what their models imply it is.

On this view, the CIPC and the PCIPC (or Matthewson's paradigm Darwinian population concept) might be viewed as starting points for reasonable heuristics for picking out populations in some cases.[17] However, satisfaction of neither definition is required for either evolution or its study. Populations with other patterns of causal interaction exist—really do exist—even when they overlap with other populations. Further, since, as I noted above, researchers must be able to *ask* of a population whether interactions relevant to evolution take place in it, there can be no specific requirement on a population that anything in particular happen in it—except that it must contain organisms. Of course, researchers will usually investigate groups of organisms in which they think it is likely that processes relevant to evolution take place, as in the HGDP research discussed above. And if one had good evidence that a population satisfied the CIPC or the PCIPC, that would make it a good candidate for study. But one nevertheless might want to study a subset of the organisms in it, as in the examples discussed both above and in the next section.

5.4 A Study of Short-Term Human Evolution

A study by Byars et al. (2010) applied well-known methods from quantitative genetics to a human population in order to argue that natural selection may be occurring over the course of only a few generations. I discuss Byars et al.'s inferences concerning natural selection in chapter 7; here I want to use the paper as an illustration of population concepts.[18]

Byars et al. (2010) investigated natural selection in a "population" (1787) of people defined by the Framingham Heart Study (FHS), a multigenerational study of cardiovascular disease among thousands of individuals (https://www.framinghamheartstudy.org). The participants in the FHS were a sample of residents of the town of Framingham, Massachusetts, their descendants, and the descendants' spouses (Dawber, Meadors, and Moore 1951; https://www.framinghamheartstudy.org/fhs-about). What is the population that Byars et al. studied? My suggestion will be that there are a few different populations, in different senses, all of which played different legitimate roles in this study. Spelling out this point will take some care. Please bear with me.

5.4.1 VARIETIES OF POPULATIONS IN THE STUDY

To start with, one possibility might be to think of the FHS participants—or at least the subsample of women that Byars et al. analyzed—as constituting a population in which biological evolution can take place.[19] This view should sound somewhat unnatural, and one of the authors of the study, Stearns (personal communication), conveyed to me that he doesn't think of the FHS participants, or any subset of them, as constituting a population in a biological sense. Rather, the women whose data was analyzed constituted a sample from a larger biological population; on this view, it would be natural selection in the larger population—perhaps the entire species—that is really of interest. Byars et al. (2010) did repeatedly refer to the people whose data they studied as a "population" (see below), but we can interpret that as a merely statistical population—that is, as the collection of individuals who are the basis of their statistical analysis—rather than as a biological population.[20]

Even putting aside issues about individuals or descendants who decided not to participate further, new spouses eventually "migrated" into the FHS "population," which was of course not causally isolated from other people in Massachusetts or the rest of the world. The environment of the FHS population also included, for example, culture and artifacts that were produced by people outside of the population per se; we can treat these as interactions with conspecifics. The FHS participants don't obviously fit any intuitive concept of biological population. Moreover, one would be interested in studying the FHS participants only as representatives of a larger population. However, it's not clear exactly what this larger population's boundaries are. (It's not the town of Framingham, since as Stearns pointed out to me, there has no doubt been migration in and out of the town since the study began, and some participants who are in the study because they are descendants of the original participants don't even live near Massachusetts.) Regardless, I don't believe that it's necessary for a determinate larger population studied by Byars et al. to be precisely defined in order for the authors' conclusions to be of value, as I'll explain. On the other hand, I think there is a way to read Byars et al's study as treating the participants in the FHS study—in particular, a subset of the women in the first three generations of the study—as the entire biological population studied.

First, although the initial FHS participants were sampled from individuals living in Framingham, Massachusetts, that sampling process isn't part of the basis for Byars et al's statistical inferences.[21] Consider these quotations:

> We measured the strength of selection, estimated genetic variation and co-variation, and predicted the response to selection *for women in the Framingham Heart Study.* . . . We found that natural selection is acting to cause slow, gradual evolutionary change. The descendants of these women are predicted to be on average slightly shorter and stouter, to have lower total cholesterol levels and systolic blood pressure, to have their first child earlier, and to reach menopause later than they would in the absence of evolution. (Byars et al. 2010, 1787; emphasis added)

Here, we report estimates of natural selection, and the potential genetic response to selection, *in the women of the first two generations of the Framingham Heart Study (FHS) population*. (Byars et al. 2010, 1787; emphasis added)

I'll discuss some of the technical terms used in these quotations in chapter 7.
Here I want to focus on the concept of population that these quotations seem
to exhibit.

First, the italicized passages state that properties said to be measured or
estimated[22] are for those women who are in the study—that is, who are part
of the "Framingham Heart Study (FHS) population." These are not estimates
concerning all residents of Framingham and their descendants, nor of some
wider group of people. They are estimates of collective properties of precisely
the women represented in the part of the data set that Byars et al. used. Second, it makes no sense to talk of two generations of the broader population
from which the individuals in the FHS were drawn. The FHS has clearly defined generations, since children of the first cohort of participants were subsequently recruited into the study, as were many of their children, and so on. The
general population of Framingham or Massachusetts or the United States has
no such discrete generations. Thus Byars et al.'s statistical inference, which assumes discrete generations, is an inference concerning a population restricted
to FHS participants. Third, Byars et al. also remark at one point that

we divided *the population* into six periods of differing average LRS by year of
birth and expressed the fitness of women relative to the average for their year
of birth. (Byars et al. 2010, 1788; emphasis added)

(LRS is lifetime reproductive success; i.e., the number of offspring that a
woman actually has. I discuss this concept of fitness further in chapter 7.) In
the context of the article, it is most reasonable to read Byars et al. as talking
here about temporal division as an operation performed on the FHS data.
Thus "the population" that was divided up consists of individuals measured
in the FHS study.[23] Further, note that at no point do Byars et al. mention the
manner by which the FHS participants were sampled from the larger population of Framingham, Massachusetts. If Byars et al. had done this, it would
suggest that the sampling methods were important to their estimates, and
that the estimates might be about this larger population. That Byars et al.
don't even allude to the sampling methods suggests that these methods are
unimportant to their statistical inferences, and in particular, to the calculation of statistical quantities such as p-values.[24]

Finally, it's not just that Byars et al. said things that suggest that one
could view the women in the FHS data as constituting a population in which

evolution takes place. The way in which Byars et al. modeled possible natural selection in their study (chapter 7) does exactly what I suggested in section 5.3.5 would be necessary in this kind of case, where there is not a real causal boundary between the population and external factors: the regression model that Byars et al. used estimates fitness relationships, but it does so in part using what are known as a phenotypic covariance matrix (P) and a selection differential vector s, which reflect both heritable influences *and* those that go beyond genetic or other heritable effects (such as cultural influences from mothers to daughters).[25] The probabilistic relationships summarized in this matrix and this vector include all factors that might influence fitness and phenotypes, including influences from the physical environment, spouses, friends, the broader culture, and other social and economic factors. That is, the way in which Byars et al. estimated collective properties such as fitnesses for the women represented in their FHS data was designed to take into account the possibility of influences from outside this group of women. Section 7.5 provides additional details.

5.4.2 COMPARISON WITH EXPERIMENTAL POPULATIONS

Treating a set of women represented in the FHS data as a biological population is just a relatively extreme case of practices that are widespread in evolutionary biology. Populations and their environments needn't be isolated from other influences, even from genetic and other interactions with members of the same species. A population-environment system can incorporate not only interactions with possible members and descendants of the original population, and possible effects of the original environmental conditions, but also causal influences that might intuitively be considered "external" to the population-environment system. What's important for scientific investigation of such a system is that researchers take into account the possibility of such influences—whether by ignoring influences whose average effects are reasonably thought to be negligible, or by including in the analysis elements that model and measure effects that are not negligible. For example, as suggested above, a model of a population might include elements that reflect measurements of migration in and out of the population (e.g., Gillespie 1998).

It will be useful to consider Samuk, Xue, and Rennision's (2018) experiment to determine whether predation selects for a change in brain size (surface area) among threespine stickleback fish. Samuk, Xue, and Rennision noted that there is already evidence that predation on fish selects for increased brain size, and that within fish species, brain size is often correlated with cognitive performance measures. Their hypothesis was that predator avoidance increased cognitive demands. However, there is also some evidence suggesting that predation might select for decreased brain size in fish. One reason for this might be, for example, that larger, more complex brains can require significantly more energy, which could perhaps be better used for escape.

Samuk, Xue, and Rennision performed experiments with threespine stickleback fish in ten experimental ponds, including five pairs of similar ponds, approximating the fishes' natural environments. One randomly chosen pond in each pair also received two cutthroat trout, natural predators of sticklebacks. A year and a half and a few generations later, Samuk, Xue, and Rennision compared brain areas of sticklebacks from the predation and no-predation ponds. Sticklebacks from ponds with predators had smaller brain areas on average. Controlling for various factors, p-values indicated that differences in the smaller brain surface areas among sticklebacks experiencing predation were significant. That is, the brain area differences would have been unlikely had they not been the result of a systematic difference between the ponds with and without the predatory trout. Samuk, Xue, and Rennision concluded that predation had selected for smaller brain size. In one section of the paper, the authors informally discussed possible lessons for thinking about natural selection on fish brain size beyond their study—not only for threespine sticklebacks but for other species of fish as well.

It's not surprising that Samuk, Xue, and Rennision repeatedly referred to the fish in the ten ponds as distinct "populations." These fish were sampled from larger natural populations, but during the experiment, members of different ponds didn't interact (thus probably satisfying the CIPC and the PCIPC). Further, evolution in each population was of course affected not only by whether trout were present but also by other conditions in each pond—other animals, plants, the physical structure of the ponds, and so on. Although the ponds in each pair were similar, Samuk, Xue, and Rennision's statistical analyses took into account the possibility that unknown differences between ponds could influence their outcomes. Thus, even though there were presumably factors in each pond environment that were relevant to the evolution of the ten stickleback populations, whatever these factors were, they were not individually represented in the analysis.

Viewing women in Byars et al's (2010) study as a population is not so different. Of course, there are numerous social, cultural, economic, physical, and other factors that affect evolution within that population, and Byars et al's analysis said nothing about those factors individually. Such factors can be understood as unknown aspects of the environment affecting the lives of the women in the study. Byars et al., however, took account of the collective influence of such factors in their analysis, just as Samuk, Xue, and Rennision

(2018) took into account unknown environmental influences in theirs. The Samuk, Xue, and Rennision study does differ from Byars et al.'s in that once the initial populations were constructed, no conspecifics from outside of the population in a pond were able to mate or otherwise interact with those in the population. By contrast, FHS participants after the first generation included people whose parents were not in the original study. (As I noted above, spouses of participants were recruited into the study, if possible.) However, we saw in section 5.3.4 that biologists do not necessarily require populations to be genetically isolated, and that investigators can deal with this complicating factor in various ways.

If I'm right that the subset of the FHS participants whose data Byars et al. analyzed can be considered a biological population in a reasonable sense, that does not negate Stearns's point that we should also view the study as investigating some broader population, from which FHS participants are a sample. Like much research in evolutionary biology, the broader goal of the research was not limited to questions about the study population. The abstract of the paper says the following:

> Our aims were to demonstrate that natural selection is operating on contemporary humans, predict future evolutionary change for specific traits with medical significance, and show that for some traits we can make short-term predictions about our future evolution. (Byars et al. 2010, 1787)

Other passages make it clear that in addition to demonstrating that natural selection affects humans on relatively short time scales, the authors wanted to illustrate some traits that might, in general, be under selection and to motivate further studies of human populations. I'll discuss further details of Byars et al.'s study in chapter 7.

5.5 What Is a Population?

What is a population? Minimally a population must be a bunch of organisms that can be treated as members of the same species in which evolution might take place. What else is required? One response to this question would be to try to define a taxonomy of population concepts, in line with some of Millstein's (2010a) suggestions. I certainly think that a taxonomy of population concepts or an investigation of how contexts disambiguate a few population concepts (or both) could help to illuminate certain sorts of cases of evolution and its study. If the approach is to try to define concepts to cover all cases, though, I worry that the proliferation of population concept definitions would never end, or that there would be important edge cases that would be handled poorly. Would it have been better if Moreno-Estrada et al. (2008, 2009) had had to come up with distinct terms for the original fifty-one populations, the thirty-nine populations constructed from them, and the seven continental populations constructed from them, all of which can be studied using the same statistical techniques? What if some other researcher grouped the same populations in a different way? How many population concepts would be needed? (Just for one data set!)

Another approach might be to use Barker and Velasco's (2013) distinction between population concepts and the assumptions that specify how the concepts are to apply in particular contexts. This means that fewer distinct population concepts would be needed, since research contexts determine different ways to apply a single Millstein-like (or Matthewson-like) population concept.[26] For example, perhaps one could say that populations like those mentioned in the quotation by Hämälä, Mattila, and Savolainen (2018) (§5.3.3) could all fall under one concept that specifies that there be certain kinds of interactions between organisms within and between populations, but that the researchers' modeling context determines whether a population would exist when there is substantial migration. On the other hand, the distinction between concepts and their contextual criteria of application seems arbitrary. What is the difference between saying that Moreno-Estrada et al. (2009) applied several different concepts to HGDP data versus saying that they used one concept but applied it using different criteria in different cases?

At this point, some readers may be wondering whether I think that nearly anything can count as a population. What about a population consisting of some members of a bird species and some members of a crab species (Matthewson 2015)?[27] Or since I am already restricting my focus in this book to populations composed of conspecifics, could a population consist of a thousand E. coli bacteria, each in its own petri dish, separated from each other by at least ten kilometers? The answer is that there's no need for population concepts to incorporate all elements that might constrain their application. Other ideas from evolutionary theory, analogies to previous research, and pragmatic factors that guide research can help scientists determine whether it is useful to treat something as a population. One still might ask, "But doesn't that mean that you *do* think that the ten-kilometer-separated bacteria could constitute a population? Nothing you've said rules that out!" Maybe, but I don't think it matters. Science doesn't need its concepts to rule out cases that no one would want to investigate anyway. And if it turned out that there *was* a good reason to study evolution in socially distanced E. coli—suppose there were evidence that bacterial fission generated quantum entanglement between distant bacteria— then the word "population" would be available for use.

Nevertheless, I do think that there is value in trying to define population concepts that capture what biologists are doing, and in explaining how these concepts could be, or are, applied in practice. Such definitions can help to clarify the nature of those research contexts within which a particular population concept applies. The philosophical research on population concepts that I've mentioned above can be seen as providing contributions toward this goal, even if they are too narrow to fit all research contexts. Nevertheless, I would discourage the assumption that the project of clarifying population concepts has the potential to capture the full spectrum of uses of ideas about populations in evolutionary biology. (For example, it seems plausible that normal natural-language style change [§4.5.3] might have led to slightly novel uses of "population" in the large literature involving statistical tests on genomic data, of which Moreno-Estrada et al. [2009] is a recent illustration. Perhaps researchers using HGDP or earlier data sets began using "population" as they did simply because it was useful to do so, and because they could see how to apply evolutionary reasoning and methods in sensible ways. Byars et al.'s [2010] research may provide another illustration of this point. But if this is how different uses of "population" develop, then trying to define *the* few specific population concepts that are supposed to capture all important uses of the term in evolutionary biology would not be useful.)

By suggesting that "population" is used in various flexible and perhaps novel ways in various research contexts, and that its uses can't necessarily be captured by a few definitions, I'm not suggesting there is no worthwhile work to do to clarify and understand the roles in scientific practice of terms such as "population." In some cases, particular, clearly defined senses of a term may be needed. I alluded above to Millstein's (2009, 2010a) remarks that a paper by two avian ecologists (Wells and Richmond 1995) described a variety of uses of "population" and advocated that uses of the term be constrained in certain ways. If, in fact, there are difficulties in understanding talk of populations between different kinds of researchers in evolutionary biology, ecology, or other areas of biology, then of course, it could be helpful for scientists or philosophers to develop standard terms for particular contexts.

However, I think that philosophers can do more than help to clarify specific concepts for specific contexts. Tracing ways that different variant population concepts derive from earlier ones, in interaction with various needs of different research contexts, might illuminate evolutionary biology and science more generally. This would be research in the style of history of science, though focused on recent science. A complementary approach that could provide something like a systematic overview is illustrated by some of Wimsatt's work (e.g., Wimsatt 2007; Wimsatt and Griesemer 2007). Wimsatt sometimes adopts a strategy of identifying various factors that influence scientific practice, with illustrations of their influence, without specifying a fixed or precise scheme for relationships between factors. This can provide insight into science when the way that science works varies in subtle ways between related research contexts, and it can allow one to apply similar ideas to new cases without expecting that they will apply in exactly the same way. One might be able do something like this for ideas that interact in biologists' decisions to use "population" in the ways they do. For example, perhaps it's possible to develop a list of heuristic rules that influence decisions about how to use "population" in practice, even if additional, context-specific factors also influence scientists' decisions.

5.6 Conclusion

After criticizing Millstein's way of defining populations, as well as others, I argued that researchers must be able to define populations flexibly, because they don't necessarily know what sorts of causal interactions the members of the populations are likely to participate in. I argued that in practice biologists do use ideas about population in flexible ways. However, that flexibility means that researchers may need to take into account differences in the likely probabilistic causal patterns in the populations they have defined. The Byars et al. (2010) example illustrates this idea, using statistical methods that can take into account interactions that might occur across the population-nonpopulation boundary. A similar point is illustrated by the quotation from Hämälä, Mattila, and Savolainen (2018) (§5.3.3), which conveyed that groups of organisms can reasonably be considered populations even when there is so much migration between them that the effects of it are much greater than the effects of natural selection. However, this implies that one would probably need to take into account migration in order to understand evolution in such a population. What is necessary in order to find evidence of evolutionary processes in not-necessarily-decomposable parts of the world is only that one take sufficient account of the ways in which the part under study is not decomposable. One must *"make sure the role of the environment is included in the model so that one can do the analysis correctly"* (Sterrett 2014, 34; italics in original). Among ways of including the role of the environment, however, we must allow both statistical modeling of unknown factors and, sometimes, simply ignoring small effects.

My view is consistent with Gannett's (2003) claim, expressed in a passage I quoted in section 5.2, that the constitution of populations varies depending on researchers' interests. I disagree with Gannett, however, that such populations are not "mind-independent objects" (Gannett 2003) in any important sense. My view is that researchers delineate a real aspect of the world about which their claims will be, if the research succeeds, approximately correct. Interests, available data, and so on can determine what the researchers choose to study, but that object of study is nevertheless real; it would be there whether the researchers studied it or not. I explore this idea in the next chapter, where I also argue that population pluralism does not threaten causalism.

Real Causation in Pragmatic
Population-Environment Systems

6.1 Introduction

In chapter 5, I argued that evolutionary biologists seem to be able to define populations in very flexible ways, and that this is beneficial to their research. I also pointed out that researchers sometimes simultaneously study evolutionary processes in an overall population and in some of its component populations. However, a number of philosophers of biology seem to believe that if there are distinct populations containing some of the same organisms, there would sometimes be no objective fact of the matter about whether selection, drift, or migration takes place, or about how strong selection is. The reason, essentially, is that selection, drift, and other processes such as migration have to be specified in terms of patterns within a population (or, on my view, a population-environment system), and the character of these patterns can vary depending on what counts as the population. Statisticalists (Matthen and Ariew 2002, 2009; Walsh 2007, 2010, 2013, 2015b; see also Stegenga 2016) have used population pluralism to argue that selection, drift, and so on are not causal factors but mere artifacts of perspectives or models. In response, causalists (Millstein 2009, 2010a, 2010b, 2014, 2015; Otsuka et al. 2011; Gildenhuys 2014) have adopted various strategies for arguing against such population pluralism.

Although my view is causalist, the statisticalists are correct up to a point. As chapter 5 argues, the same organisms *can* be members of multiple populations that are specified by different research or modeling choices. As a result, the same organisms can be part of processes undergoing natural selection in different directions, with different strengths, or they can be part of different populations undergoing selection on one hand or pure drift on the other. I differ from statisticalists in that I think it can be correct to view selection and drift as real causal factors affecting the evolution of populations in different, overlapping population-environment systems.[1] My view is that causes that are, loosely speaking, the "same"—selection for trait A over trait B, for example—are not actually the same causes when they are treated as causes in different population-environment systems. Roughly speaking, the causal probabilities that control the evolution of a population, and that are the basis of natural selection and drift (etc.), are *analogous to* probabilities conditional on a population and environment (§2.3, §2.2.2). However, there is nothing contradictory about an outcome having different probabilities conditional on different events: $P(A|B)$ need not equal $P(A|C)$. Similarly, there is nothing contradictory about the "same" outcomes having different causal probabilities in different population-environment systems, even when the systems overlap physically or temporally (as I explain below).

Some provisos: My conclusions concern the causal structure of evolution and its relationship to different ways of studying evolutionary processes, some of which were illustrated in chapter 5. Though I'll frame some arguments as responses to Denis M. Walsh and other statisticalists, my conclusions should be of interest to readers not particularly interested in statisticalist/causalist debates. And while I will use fictitious cases to make ideas clear, the examples are inspired by real-world phenomena, and my conclusions are applicable to real-world empirical research and modeling. (When appropriate, I'll make it clear how my arguments connect to biological research described more fully elsewhere in the book.) It will simplify my presentation if I make the points here in terms of different kinds of selection, without discussing drift in a significant way. Nevertheless, some readers will want some indication of what my view is about drift, so I'll briefly sketch a view at the end of the chapter. The views that I advocate here are also relevant to migration and mutation, even though I won't discuss these evolutionary factors either. Also note that in this chapter I treat causal type fitness as defined in specific ways that should be clear from context but, consistent with my remarks in section 4.5, I make no pretense at any point that I have defined a canonical conception of causal type fitness applicable to all cases.

In section 6.2, I discuss ways that what counts as an environment depends on the population that is under investigation. In section 6.3, I argue that the fact that natural selection can have different strengths and directions for populations that overlap with each other is consistent with the view that natural selection is a cause of evolution. I make a similar point in section 6.4 about cases in which environmental variations can be specified as either the result of chance processes or taken as given after the fact. Section 6.5 briefly summarizes my views about the relationship between drift and selection in the context of the issues discussed in this chapter.

6.2 The Problem of the Reference Environment

In chapter 4, I gave general principles for how subenvironments that are part of a whole environment affect evolution of a population. What determines the whole environment, though? In an earlier work, I called this "the problem of the reference environment,"[2] and I suggested that the whole environment includes all "conditions relevant to reproductive success for at least some members of a population" (Abrams 2009c, 24). This idea is similar to Brandon's ecological environment concept (§4.2.1). I made it clear that relevance to reproductive success depends on a time period over which evolution might take place, since different environmental conditions might come into play over different periods (see chapter 4, especially §4.3). Thus by choosing to study a particular a period of time over which evolution takes place, and specifying the individuals that make up the population to be studied over that period time, one implicitly specifies a range of environmental variation that matters for the evolution of that population. For example, a population evolving over a hundred years might experience a different range of environmental variation than the same population over a single year. Fitnesses affecting the probable changes in frequencies of the same set of types may be different depending on whether research questions concern one period of time or another.

A different principle of relativity of environments to populations was independently given a few years later by Millstein: "the boundaries of the environment would be delineated by the fullest extent of the spatial location of the population" (2014, 749). It's possible that Millstein (2014) intended her "spatial extent" specification of the whole environment to imply something like my "conditions relevant to," but if so that implication wasn't clear. I think my characterization is preferable for two reasons. First, not everything occurring in a spatial region must be relevant to the evolution of a population (§4.2.1). There may be some variations that make no difference to natural selection on a given population. For example, whether a particular small clump of dirt is tilted slightly north or south might make no difference to fitness in a population of grazing animals. Thus some conditions within a spatial extent, or variations in them, need not count as aspects of the environment relevant to evolution. Second, once we treat the environment as defined by conditions relevant to evolution, spatial boundaries of relevant conditions can become very complex in ways that don't really matter to our understanding of evolution. Whether or not there are true "butterfly effects" (Lorenz 1993)—in which tiny variations in conditions on one side of the earth can affect conditions on the other side—evolution can certainly be affected by weather conditions, which can depend on influences from distant regions. So my "conditions relevant to reproductive success for at least some members of a population" seems like a better starting point for a way of specifying what counts as an environment relevant to evolution.

However, as previous chapters make clear, what happens in population-environment systems is generally chancy. So what counts as "relevant to reproductive success for at least some members of a population" will depend on chances of environmental conditions occurring—and occurring in ways that affect outcomes for particular realizations of organism traits (in organisms). Consider for example, a population of insects living in a desert region. If there is a chance that rain will occur in this region, but the chance is very small, say 0.014, *rain with chance 0.014* can be considered an aspect of the environment. This can be considered a different environment than one in which the chance of rain is 0.64. (Alternatively, one might say that it's wrong to think that environmental conditions are simply relevant or not. Rather, relevance to evolution is a matter of degree, measured partly by probabilities.) Note further that the chances of conditions affecting population members may be different during different periods of time. Environments can reflect, among other things, both periodic changes in probabilities of conditions occurring, as with seasonal weather cycles, and progressive changes, as in the effects of climate change (see §6.4). Also implicit in this picture, based on the population-environment system conception described in chapter 1, is that chances may be different relative to different starting points for a population-environment system. (Note that one dimension of chanciness in this kind of conception of environment is that what counts as a member of a population, after a population-environment system begins, can also be chancy, as noted in chapter 5. The probabilities that matter are probabilities of interactions involving organisms that realize various possible inheritable traits.)

The preceding characterization of the environment as consisting of conditions that may, with particular probabilities, affect the evolution of the population is intentionally vague. In principle, that butterfly on the other side of the planet can be treated as part of an environment, but more usually the environment would be conceptualized by researchers in terms of likely effects of those butterflies in conditions that are more closely related to organisms in the population under study. In practice, researchers might decide that some environmental conditions should be treated as irrelevant, even if the researchers think that these conditions have some chance of influencing the evolution of a population (chapter 5). For example, if the chance of rain is thought to be 0.014, this might be modeled as no rain at all—but that kind of approximation is routine in modeling. Thus pragmatic research concerns can determine whether to take certain improbable environmental variations seriously. More generally, as I also explain below (see also chapters 5, 7), one can treat one set of conditions, or even a spatial region, as "the environment," while either taking into account possible patterns of causal interventions on that environment or treating them as sufficiently negligible to ignore. My arguments suggest that scientists can use both "population" and "environment" in a variety of ways that nevertheless capture real causal facts about evolution in real populations and environments, so defined, *relative to* those characterizations of a population and environment. Spelling out these ideas is one of the main purposes of this chapter.

6.3 Intersecting Population-Environment Systems

In this part of the paper, I use fictitious, simplified examples to help clarify philosophical issues that seem to be raised by scientists' flexible uses of "population" discussed in chapter 5.[3]

6.3.1 THE PIGEON ILLUSTRATION

Imagine a population of feral pigeons that live in a coastal area with a sandy beach below rocky cliffs. These pigeons have a set of alternative traits:[4] Pigeons with the *dark* trait are mostly dark, almost black, while pigeons with the *light* trait are mostly light gray, almost white. Thus, suppose that in an area with dark sand, *dark* pigeons are less noticeable to hawks than *light* pigeons, while in an area with light sand, *light* pigeons are less noticeable.[5] (For an analogous real-world example, see Barrett et al. [2019]. There researchers introduced barriers into a natural environment to constrain deer mice to areas with dirt of different colors and found resulting selection on fur color due to predation by birds.)

Assume that the overall population of pigeons is made up of two subpopulations P_D and P_L (D for "dark" and L for "light"), living respectively in environments E_D and E_L. Environment E_D has mostly dark sand, while environment E_L has mostly light sand. (Note that the subscripts in the names P_D and P_L label these supopulations' environments; both P_D and P_L include pigeons that are *dark* and pigeons that are *light*.) Because of predation by hawks, the *dark* trait is fitter than the *light* trait in E_D, while *light* is fitter in E_L. If there were interbreeding or migration between the two subpopulations, both environments might be relevant to the long-term fitness of pigeons in each subpopulation. Let's assume, though, that probabilities of interpopulation interactions such as migration and interbreeding are so low that there is unlikely to be any interaction over the period of time of interest to us. The two subpopulations are thus populations according to the probabilistic causal interactionist population concept (PCIPC) defined in chapter 5, and they are likely to remain populations according to Millstein's causal interactionist population concept (CIPC) described there as well. Satisfying these criteria for populationhood will allow me to present arguments for conclusions that might antecedently seem implausible. I will nevertheless refer to all of the pigeons together as a population, P_{both}.

Assume that the subpopulation sizes are always equal, that the initial frequency of *dark* in E_D equals the frequency of *light* in E_L, and that the values of fitnesses in one environment are reversed in the other environment: the fitness of *dark* pigeons in E_D is equal to the fitness of *light* pigeons in E_L, and vice versa. Then the fitness relations in the populations are mirror images of each other; probabilities of increase in frequency of *dark* in P_D will be equal to probabilities of increase in frequency of *light* in P_L. Fitnesses of the two types in the combined population $P_{both} = P_D \cup P_L$ are thus equal.[6]

6.3.2 POPULATION PLURALISM

Question: Is there natural selection on the subpopulation P_L found in environment E_L? Modeled as members of subpopulation P_L alone, there would seem to be selection for *light* among those pigeons. However, if we treat these same pigeons as members of the combined population $P_{both} = P_D \cup P_L$, there seems to be no selection for *light* overall. It thus appears that merely by treating pigeons as members of a larger or smaller population, we can make it the case that selection does or does not occur. In a similar context, Denis M. Walsh (2007) argued that since, in such cases, whether selection occurs depends on how the population is described, natural selection is not a cause of evolution. The idea is that whether a cause exists is a fact about the world, not about how we describe it, so it shouldn't depend on how a situation is described. So population pluralism seems to imply that natural selection is not a kind of cause of evolution. On the other hand, since conditions in environments E_L and E_D don't influence outcomes for the population living in the other environment, it might seem reasonable to view the situation as one in which there is selection for *light* in E_L and for *dark* in E_D. On this view, what happens to P_{both} in the overall environment is not really natural selection as a distinct causal process. This is what a view of populations like Millstein's CIPC implies: there is no interaction between members of P_D and P_L, so P_{both} is not a real population. If P_{both} is not a population, natural selection cannot take place in it. In summary, according to Walsh's argument, if P_{both} is a population, then natural selection is not a cause. If natural selection *is* a cause, it must be that P_{both} is not a population. The CIPC gives an account of *why* P_{both} is not a population. If we replace the italicized "no interaction" above with "no probability of interaction," the PCIPC can do the same job.

I've expressed the preceding argument against population pluralism in terms of a toy model, but similar arguments might be made about real empirical research. It's entirely possible, for example, that within the East Asian continental group population used in Moreno-Estrada et al.'s (2009) analyses (chapter 5), there was almost no interbreeding between some neighboring smaller populations (over the past millennia relevant to the authors' statistical analyses), thus providing a version of the assumptions in the pigeon example.

6.3.3 CAUSAL RELATIONS

On my view, however, the reason that selection seems to depend on whether members of P_L are described as members of P_L or as members of P_{both} is just that the descriptions pick out different causal relations, involving both different effects and different causes.[7] When we ask whether there is selection in P_L, we are asking about causes of a particular property—in particular, about causes of a certain kind of change (or lack of change) in frequencies of traits that might be realized in P_L. We can also ask about selection in the larger population P_{both}, but that is to ask about change of frequencies in a different population, and hence about the cause of a different evolutionary effect. These are, from my point of view, processes in different population-environment systems. Just as a change in the composition of P_L is a different effect than change in P_{both} (as a whole), selection in P_L is not the same cause as selection in P_{both}. One involves processes in one part of the world; the other involves those processes and other processes too. Of course, processes in P_{both} supervene on processes in its subpopulations; nevertheless, whether there is selection in P_L is not the same as whether there is selection in P_{both}. In terms of population-environment systems that determine causal probabilities of various possible changes in trait frequencies, the point is that distinctly specified population-environment systems can generate different probabilities for outcomes that can be described in similar ways. We can understand many different ways of carving up the world into populations and their environments as specifying real portions of the world, as long as the causal relations relevant to what probably happens in those portions of the world are appropriately understood and modeled (see chapters 5 and 7 for more discussion). Thus there is no contradiction in saying that there is selection for *light* in P_L but no selection for *light* in P_{both}. In section 6.1, I said that this point is analogous to the fact that it's not contradictory for the same outcome to have different conditional probabilities for different conditioning events. The reason that it's merely analogous to this standard point is that a population-environment system is not a conditioning event in a conditional probability (§2.3, §2.2.2).

My suspicion is that some others have not seen this point or have not adopted it because of a mistaken focus on actual token organisms. Statistical-ists and causalists alike may have thought that if we are free to define different populations containing some of the same token organisms, there would be no fact of the matter about the strength and direction of selection on those shared organisms. However, my arguments in chapters 3 and 4 show that it's a mistake to think of natural selection as acting on actual token organisms. That means that we don't have to worry that a single organism or a set of ac-tual organisms could be subject to different sorts of selection depending on the population to which we assign it or them. There are no selective forces on token organisms per se. Natural selection is always relative to a population as a whole, or more properly to a population-environment system.

(Note that on the assumption that natural selection acts on token organ-isms individually, a critic of population pluralism might argue that since no particular organism in P_L is likely to experience environment E_D's dark sand, E_D can't be relevant to selection on members of P_L. A similar argument could be made concerning members of P_D with respect to environment E_L. Thus, the argument would go, it makes no sense to talk about natural selection on members of P_{both} in E_{both} since members of P_{both} don't experience a common environment. If that view were correct, however, then for any organism o_i in P_L, we should make the same argument about those subparts of environ-ment E_L that would have no influence on what happens to o_i. After all, there is presumably natural selection on other properties of these pigeons, but the individualized force of selection that token organisms experience—I am not sure how to make sense of this idea[8]—would only apply in certain environ-mental circumstances [chapter 3] in, say, E_L. Thus much of E_L might be irrel-evant to natural selection on many individuals in P_L. Only the circumstances experienced by each particular organism would be relevant to selection on it. In effect, each organism would be in its own environment. This picture is far from anything in evolutionary biology, but cf. section 4.4.)

6.4 Environmental Change over Time

In section 6.3, I argued that natural selection can be a cause of evolution even when it has different directions or strengths because synchronic environ-mental variation allows overlapping populations to define different sorts of environments. Denis M. Walsh (2013) argued that different ways of describing *diachronic* environmental variation can determine whether natural selection or drift takes place. Given my focus on selection, I'll make Walsh's points using arguments showing that natural selection can have different directions when diachronic environmental variation is described in different ways. Walsh's argument, again, is that the character of causes shouldn't depend on mere description, so if natural selection varies in response to descriptions, it must not be a kind of cause. My response to this sort of argument will be similar to my response above: by specifying diachronic environmental variation in different ways, we can pick out different population-environment systems that incorporate some of the same stuff in the world. However, we thereby pick out different real effects and causes. As we'll see, the issues here are more subtle than in preceding sections. I'll start with an empirical illustration, but it will be helpful to use toy models to clarify the issues.

6.4.1 STOCHASTIC ENVIRONMENTAL CHANGE

Hamann, Weis, and Franks (2018) studied responses to drought in two populations of field mustard plants (*Brassica rapa*) in Southern California, over an eighteen-year period from 1997 through 2014.[9] The researchers were interested in the extent to which drought conditions during the study period resulted in natural selection for responses to the resulting reduced water availability during winter (the plant's growing season). Hamann, Weis, and Franks used seeds that had been collected and preserved from four years (1997, 2004, 2011, and 2014) over the study period, and they grew them in a variety of different, controlled conditions in a greenhouse. Among other things, the researchers found *some* evidence that there had been natural selection for early flowering as a result of drought. This was thought to be an adaptation for the shorter growing season produced by increased drought conditions.

What I want to focus on here is the fact that Hamann, Weis, and Franks (2018) viewed their study as investigating the field mustard populations' responses to the particular sequence of weather conditions that occurred during the 1997–2014 period. For example, plants collected in 2004, which followed a series of dry years, flowered earlier on average than plants collected in 1997. Plants collected in 2011, after a couple of wet years, flowered earlier than the 1997 and 2004 plants, however—in conflict with the hypothesis that drought would select for early flowering. Plants collected in 2014, after another series of dry years, flowered earliest of all.[10]

This was a study of a particular, concrete sequence of weather conditions in two nearby environments, rather than a study of wet and dry conditions in general. However, the authors also viewed this sequence of weather conditions as stochastic effects of climate change. The authors explicitly describe their research as a study of effects of climate change, but note that

> climate change has not proceeded at a steady rate. Variability between and within years has been and will remain the rule, and in fact may even increase. (Hamann, Weis, and Franks 2018, 2683)

The authors describe this variability as stochastic:

> While the general trend toward increasing severity of drought creates important selective pressures, the stochastic occurrence of wet seasons creates counterproductive (over the long-term) selective spells. (Hamann, Weis, and Franks 2018, 2692)

As this was a paper on plant evolution rather than, say, philosophy of probability, it's unremarkable that the paper doesn't clarify the sense in which the occurrence of rainy or arid years is to be considered "stochastic." Nevertheless, what Hamann, Weis, and Franks wrote is consistent with my view that one should understand population-environment systems as incorporating chances of environmental variation over time. From this point of view, we can interpret Hamann, Weis, and Franks's study as a partial illustration of the following ideas:

1. At an initial time (e.g., mid-1997) there are various chances that certain alternative environmental conditions will be realized at various times. These chances may be different for different future times, as they would be, for example, if the chance of arid winters in Southern California increased each winter due to climate change.
2. The chances that contribute to fitness differences and natural selection reflect these chances of environmental variations.
3. One can nevertheless study a particular realization of such (chancy) environmental conditions as they in fact have varied over time, investigating how that realization of environmental conditions affects the chances that contribute to natural selection at different times.

The latter is what Hamann, Weis, and Franks (2018) did, on my view. They studied ways in which the actual sequence of environmental conditions contributed to natural selection in the two field mustard populations. That one can do this is consistent with the idea that these actual environmental states are among many conditions that the population-environment system could have realized with various probabilities.[11] The following sections help to clarify this idea, as well as some of the points made in chapter 5.

6.4.2 THE INSECT ILLUSTRATION

Denis M. Walsh (2013) and I (Abrams 2009c) discussed (similar) hypothetical examples involving change over several generations. Walsh used the examples to raise potential challenges to the view that natural selection or drift are causal factors contributing to evolution. I'll use variants of my own examples to address Walsh's arguments below.[12] These simple toy examples will help to clarify the nature of population-environment systems in examples such as Hamann, Weis, and Franks (2018) and Byars et al. (2010). The complexities of real examples are important for some situations but would interfere with the presentation of new ideas in this case.

Consider a population consisting of members of an insect species with haploid genetics, which reproduce asexually during a short period of time and then immediately die. Assume the insects are living in a more or less arid region without significant seasonal weather variation, but with occasional rain. The soil in this area lies on top of a layer of rock several feet below the surface. At time t_0, each member of this population has one of two alternative phenotypes, which I'll call "*deep*" and "*shallow*," determined by alleles at a single locus. The difference between the *deep* and *shallow* phenotypes is that *deep* insects tend to spend more time farther underground than do *shallow* ones, although insects of both types routinely move back and forth between the surface and the rock below. During the usual dry conditions, there is some moisture several feet below the surface; this moisture tends to help the *deep* insects produce more offspring than *shallows* do. On the other hand, when it occasionally rains, water collects in the areas favored by *deeps*, causing some of them to drown. I will also assume that the population is not so small that the effects of drift due to randomly caused deaths are likely to overwhelm the effects of natural selection described below.

Which type, *deep* or *shallow*, has greater probabilities of having more descendants? During a wet period, *shallow* might have a greater probability of short-term reproductive success, since *deeps* are more likely to drown then. Yet because *deeps* are more likely to have more offspring during dry periods, one can easily imagine that over a longer interval, the *deep* type might have a greater probability of success. Insects can reproduce and die on a very short time scale. Is rain relevant to natural selection on this population if it is unlikely to rain for another six months? For a few years (see §4.3)? For example, suppose for the sake of a simple illustration that insects of each given type always have the same number of offspring in each of two kinds of environments, *dry* and *wet*, as specified in table 6.1.

TABLE 6.1: Offspring per individual per generation for
traits *deep* and *shallow* in two environments, *dry* and *wet*

	Dry	Wet
Deep	2	1
Shallow	1	2

If the chance of rain is negligible over a period of time, the number of *deeps* will nearly double every generation, while the number of *shallows* will remain the same. (Analogous if less extreme results would obtain with more realistic fitnesses.) If, in addition, there were resource limitations or predators that randomly remove individuals from the population, *shallow* might go extinct. If the probability of rain is low but not too small, the frequency of *shallow* is still likely to decrease, but not as quickly. For example, if rain can either be present or absent for an entire generational period, and the chance of rain during such a period is 0.25, then the probability-weighted average number of offspring for an individual of each type is

$$deep: \quad 2 \times 0.75 + 1 \times 0.25 = 1.75$$
$$shallow: 1 \times 0.25 + 2 \times 0.75 = 1.25$$

Thus in an average generation, the frequency of *deep* will increase 1.75/1.25 = 1.4 times that of *shallow*. This is one sense in which *deep* can be considered fitter than *shallow*, and it's likely that the *deeps* will increase in frequency over a long period of time. Nevertheless, over the course of a single *wet* generation, *shallow* is fitter.

6.4.3 SHORT-TERM VERSUS LONG-TERM ENVIRONMENTS

The remarks in the preceding section can be summarized by saying that *shallow* is fitter than *deep* in a wet generation, but *deep* is fitter over a large number of generations because of the probable rarity of rain. That is, there is a greater causal probability that *shallow* will increase its frequency more in a single generation that is wet, but there is a greater causal probability that *deep* will increase in frequency over many generations. Denis M. Walsh (2013) viewed this kind of difference as implying that natural selection (and drift—but I focus on selection) is description-relative and hence not causal. If we describe the population as evolving over a single anomalous year in which there is a lot of rain—or rather, in which a lot of rain is very probable—then there is selection for *shallow*. If instead we describe the population as evolving over a

one-hundred-year period in which rain is probably rare, there is selection for
deep. Thus the direction of natural selection depends on how a population is
described. Walsh would therefore argue that this means that natural selection
is not a causal factor that contributes to evolution.[13]

However, as I have argued above, fitness and selection are relative to a
specification of a population, as well as to an interval of time (§4.3). These
determine probabilities of encounters with various potential environmental
conditions. Selection can have different strengths and directions over different intervals of time, because probabilities of evolutionarily relevant events
differ over those intervals. Nothing about this implies that we are not talking
about causal influences. They are simply different causal influences, some of
which supervene on others. This is so even though—or rather because—a
wet environment in one year plays a role in the influence produced during a
sequence of years including that year. In a more realistic case, the dimensions
of environmental variation would be greater, but the same point would hold:
the probable environmental variations differ for different periods of time, and
thus the causes of evolution differ as well.

6.4.4 RELATIVIZING TO ENVIRONMENTAL SEQUENCES

Denis M. Walsh (2013) also argued that selection and drift are description-
relative and not causal because their influence seems to depend on whether
we describe evolution in terms of probabilities of environmental changes or
in terms of actual environmental changes. This case is quite similar to the
preceding one, but it brings out points that might not yet be clear. Parallels
to my brief remarks on Hamann, Weis, and Franks's (2018) study above will
be apparent. I'll again convey Walsh's main idea in terms of different kinds of
selection, leaving discussion of drift for section 6.5.

Assume there is an initial population of one thousand *deeps* and one thousand *shallows* that can grow without restriction. The *wet*, rainy environment
has a probability of 0.25, and the *dry* environment has a probability of 0.75.
Fitnesses are as in table 6.1, and other parameters are as described above.
Since in this model all individuals of a given type have the same number of
offspring in a particular environment, population-wide environmental variation is the only stochastic effect. Suppose that the actual sequence of environments experienced by the population is *wet, wet, wet, dry, wet, wet, dry, wet*.[14]
This sequence is quite improbable, since 75% of the environments in the sequence are *wet*, but *wet* has only a 0.25 probability.[15] However, improbable
things do occur in nature. As table 6.2 shows, with this particular sequence
of environments, the number of *shallow* individuals would be sixteen times

TABLE 6.2: Evolution in the insect example with a particular sequence of environments

Environment	Offspring per individual		Number		Generation
	Deep	Shallow	Deep	Shallow	
Wet	1	2	1000	1000	1
Wet	1	2	1000	2000	2
Wet	1	2	1000	4000	3
Dry	2	1	1000	8000	4
Wet	1	2	2000	8000	5
Wet	1	2	2000	16,000	6
Dry	2	1	2000	32,000	7
Wet	1	2	4000	32,000	8
			4000	64,000	9

Note: The numbers of *deeps* or *shallows* listed in each row are those in the population at the beginning of the period of time for the generation. After the first row, these numbers are products of the numbers in the previous row with fitnesses (offspring per individual) listed in the previous row. The last row gives the numbers of *deeps* and *shallows* after generation 8, at the beginning of generation 9. (Since the model doesn't continue at this point, there is no need for entries in the first three columns for generation 9.)

the number of *deep* individuals at the beginning of the ninth generation. (The same point could be made with a more realistic example by changing fitness values, increasing the population size, allowing more variation in fitness, allowing more variation in environmental conditions, etc.)

Now, although in this particular (improbable) sequence of environments, *shallow* greatly outreproduces *deep*, if instead we average over *all possible* sequences of *wet* and *dry* environments, weighted by the probabilities of each sequence, we find that *deep* outreproduces *shallow* on average. The average number of *deeps* in the ninth generation is about fifteen times the number of *shallows*, as table 6.3 shows.[16] Thus if we describe the evolution of the population purely in terms of average long-term outcomes beginning from the initial state, it appears that selection favors *deep*. However, if the sequence of environments is the particular one described—which is certainly consistent with the setup—it looks like selection favors *shallow*. So, selection can be described in two different ways. Walsh would again argue that this means that natural selection cannot be considered a cause of evolution. Note that Hamann, Weis, and Franks (2018) examined evolution on the assumption that a sequence of environments was already fixed, but they also thought that this sequence was the result of an ongoing stochastic process. Thus Walsh's (2013) argument may seem applicable to their research, in which case natural selection per se should not be considered a cause of the evolution of field mustard in California in recent years.

TABLE 6.3: Evolution in the insect example: Average numbers of descendants

Number		
Deep	Shallow	Generation
1000	1000	1
1750	1250	2
3063	1563	3
5359	1953	4
9379	2441	5
16,413	3052	6
28,723	3815	7
50,265	4768	8
87,964	5960	9

Note: The number of individuals of each type in each generation is calculated by averaging over numbers of individuals in all possible sequences of dry and wet environments.

Here is my characterization of the general form of Walsh's argument:

We can describe the effects (e.g., reproductive events) of a chancy sequence of events (environmental states) either in terms of the probabilities of all such sequences, or in terms of the actual sequence that occurs. Since the effects would be different (or would have different chances) depending on which description we chose, the events (environmental states) don't cause the effects.

Let's consider applying similar reasoning to a betting game in which two coins are tossed five times each, and a gambler's winnings depends on the sequence of outcomes in pairs of tosses. (For example, a gambler might bet on whether the coins are both heads or both tails.) Suppose though that the croupier only has one coin. She tosses it five times, records the sequence of values, and then tosses it five times again, pairing each outcome in sequence 1 with its corresponding outcome in sequence 2 to create a sequence of pairs. One way to describe this is as a sequence of pairs of independently determined coin tosses. Another way to describe it is to treat the first sequence of tosses as fixed when the second sequence is performed. Conditional on the states from the first sequence of tosses, the probabilities of pair outcomes are different from what they would be considering the two sequences as the result of chance. For example, an outcome of two heads in the first pair has probability 1/4 if the earlier sequence is not treated as fixed, but it has probability 1/2 if the earlier sequence has already assigned heads to the first toss in the first sequence and we conditionalize on the assumption that the first toss comes up heads. Walsh's argument implies, it seems, that the tosses did not cause the sequence of outcomes, since the same sequence of tosses can be described in different ways. Notice that the fact that one coin was tossed earlier and then later is completely inessential to the example. The same argument could be made for two coins tossed simultaneously, by treating the actual outcomes of one coin as given by what actually occurs, while treating outcomes for the other coin as subject to chance. The fact that Walsh's argument concerned temporal sequences of events is not essential to the argument. More generally, Walsh's strategy, if legitimate, has the potential to show that there is a lack of causation in any system that is (a) chancy and (b) has multiple simultaneous states, since then we can derive different values for the chances of some states depending on whether we describe other states as having the fixed values that they actually took on

If this kind of argument is valid, how far can it be generalized? Could this kind of argument be applied to the chancy ways in which molecules move and interact within cells, by considering molecular states given other states treated as given? Then, since probabilities of interactions would differ depending on how the situation was described, it may be that cellular behavior could not be caused by such interactions. Further, since behavior of an organism as a whole depends partly on behaviors and interactions of its cells, do we then have to say that behaviors of organisms are not caused by anything internal to the organisms? It's worth emphasizing at this point that Walsh and other statisticalists have repeatedly suggested that they think that actions of token organisms do cause evolution (Matthen and Ariew 2002, 2009; Walsh 2015a; Walsh, Ariew, and Matthen 2017). Walsh's argument strategy threatens to undermine this assumption.

My own response to Walsh's argument will seem familiar by now: there are many kinds of coinstantiated or overlapping real population-environment systems; the questions we ask can pick out one rather than another. To ask about the evolution of a population *assuming* a particular sequence of environments is to ask about a different kind of causal process in the world than to ask about the evolution of a population in an environment in which alternative environments are subject to chance. It is to ask about different chance setups. (Similarly, to ask about the evolution of a gambler's wealth on coin-toss-pair bets assuming an actual sequence of outcomes for one coin is different from asking about a system in which those same outcomes are treated as chancy.) Even knowing how the environmental variation has played out, there are facts of the matter about what the chances of the environmental states were, and we can ask about evolution in such a chancy environment, rather than relative to the actual sequence of environmental states.[17] Again, researchers' interests allow them to pick out various closely related aspects of the world that are real, despite their pragmatic specification. In doing so, they pick out population-environment systems. This conception is broader than the initial conception of a population-environment system introduced in section 1.4.3, allowing population-environment systems to be specified as constrained by known (or estimated, or likely) conditions. A population-environment system is still a chance setup, just as a gambling game with pre-specified outcomes on some coin tosses is a chance setup.

This view, in addition to making sense of researchers' claims in cases like Hamann, Weis, and Franks (2018), is consistent with modeling practices in evolutionary biology. For example, one modeling strategy for studying the time that it takes an allele to reach fixation[18] involves calculating probabilities of times to fixation on the assumption that the allele does reach fixation (e.g., Waxman 2009). Or consider studies of real-world populations using coalescent simulations (e.g., Voight et al. 2006; see also Hein, Schierup, and Wiuf 2005; Wakeley 2009). These start from certain characteristics of contemporary genetic data and generate possible ancestral lineage trees in reverse time order in a way that is consistent with characteristics of the current data.[19] Such models are like the fixed environment sequence modeled above in that they examine evolution on the assumption that later states take on certain values. So modeling evolution on the assumption that certain later states are fixed seems to be a legitimate part of scientific practice. Researchers who use such models clearly also assume that there is a probabilistic process that starts from an initial state and moves forward in time. (Coalescent models are consistent with the population-environment view if we treat the calculation of probabilities of earlier states given later ones as derived from a population-environment system that begins at an even earlier time. This earlier time need not be specified, if relevant earlier times can be assumed to determine roughly the same probabilities of later outcomes.)

6.4.5 CHANCE AND ENVIRONMENTAL SEQUENCES

It may be helpful for some readers if we look ahead to chapters 8 and 9 in order to clarify the sense in which a population-environment system can be specified by a fixed sequence of states. I indicated in the previous section that a population-environment system with a fixed sequence of environmental states is consistent with, and overlaps with, a different population-environment system, beginning from the same (token) initial state, in which the fixed sequence is only one of many possible sequences that have a chance of being realized. There are two senses in which this is true, though only one of them is what I intended above, and it is the only sense that's relevant to evolutionary biology as such.

1. A population-environment system with a fixed sequence of environmental states could exist because it just happens to be the actual way that the environments got realized. This claim is compatible with chances of state sequence realization that are *fundamentally* chancy—as perhaps quantum mechanical propensities would be. Now, I don't believe that fundamental propensities play a significant role in population-environment systems (§8.3.2), and chapters 8 and 9 explore some ways that a population-environment system can incorporate chances in a manner that is consistent with an underlying deterministic development of the events that realize a population-environment system. On such accounts of population-environment system chances, there is a sense in which the development of a population-environment system is chancy, while in another sense—at another level, I'll say—it must unfold in the way that it does. (This is the first sense in which a sequence of environments might be fixed.) If this view is correct—if there are population-environment system chances of a kind that makes them consistent with underlying determinism—then I believe we can still think of a population-environment system as chancy, and as such, there will in fact be an environmental sequence that is the way that the system happened to develop consistent with the chances—as above. And we can still think of the fixed environmental sequences as constraints on chances when a researcher specifies a population-environment system in terms of such constraints, as in the preceding section.

2. However, if population-environment systems are realized by processes that are close to deterministic, as I believe they are, then there is a (second) sense in which there is nearly always a fixed environmental sequence—and a fixed everything else that happens in a population-environment system. At this level, it's always the case that a population-environment system must unfold exactly as it does (apart from unusual quantum mechanical events). Seeing population-environment systems in this way puts to one side the notion that the events that realize a population-environment system are chancy. As a result, this view does nothing to help us understand the success of evolutionary biology on its own terms. Evolutionary biology is, after all, in large part a science of probabilistic relationships (as illustrated throughout this book).

I think it's important to try to understand the basis of the causal probabilities (chapters 2, 7) that population-environment systems realize, which is the subject of chapters 8 and 9. And I believe this basis involves underlying determinism. However, treating this underlying determinism as itself the source of a fixed sequence of environmental sequences (that is, the second sense) is not directly relevant to understanding the ways in which research in evolutionary biology is carried out. The first way of thinking about fixed environmental sequences is.

6.5 Drift

For interested readers, I'll sketch my current view of drift, extending the above discussion of selection while drawing on the view of population-environment systems described in chapter 1 and section 4.2.3, with inspiration from Wright-Fisher models (§0.3.3, §0.3.4). This brief sketch may not satisfy everyone, but I want to at least point in the direction of a view compatible with the current framework. There is some additional discussion in Abrams (2007b) (though I no longer endorse everything in it). In a paper in preparation (Abrams Drift MS), I criticize other philosophical views of drift and develop the present view in more detail. Clatterbuck, Sober, and Lewontin (2013) have endorsed a view similar to mine, and papers by Huneman (2015) and Strevens (2016) include related ideas.

My view is that selection and drift as causal factors (see §0.3.4) are not distinct processes or distinct causes. They are aspects of a single process of evolution for a population-environment system, or more specifically, they are aspects of a causal probability distribution over future states of the evolution of the population.[20] The shape of the probability distribution, however, usually varies with the size of the population in the population-environment system;[21] this variation in the shape of distributions is what constitutes drift as a causal factor. The distribution also varies with fitness differences, but in different respects, which constitute natural selection as a causal factor. Thus, roughly speaking, drift is the aspect of this probability distribution that can be manipulated by modifying population size, and natural selection is the aspect that can be manipulated by modifying causal type fitnesses (see figures 0.1 and 0.2 in chapter 1).

Earlier in this chapter, we saw that different ways of carving up a population give rise to different causal probability distributions over future states of the (different, overlapping) populations. I illustrated this idea in terms of different directions or strengths of natural selection, but the same point applies to drift: other things being equal, it's more probable in a small population that frequencies of outcomes will shift in ways that don't depend on fitness differences. This difference in the strength of drift is entirely consistent with, for example, a smaller population being a subset of a larger one. In the larger population, the fact that there are more organisms means that the per-organism variances in outcomes are more likely to cancel out. That is, when there are many organisms—many trait realizations—the fact that one organism has more offspring is likely to be counteracted by the fact that some other organism of the same type has fewer offspring. This is not a claim about collections of token organisms, though. It's an illustration of a point about chance setups, and more generally about probability: increasing the number of possible trait realizations by increasing initial population size changes the causal probability distribution over ways that frequencies of traits in a population-environment system might evolve. (The coin example in §0.3.4 illustrates the same idea.)

The relativity of the strength of drift to the choice of population has been used by statisticalists to argue that drift is not a cause (e.g., Matthen and Ariew 2002, 2009; Walsh 2007, 2010, 2013, 2015b). My response to those arguments, though, is exactly like responses earlier in the chapter: drift as a causal factor is relative to a specified population. The fact that one population is a subset of another doesn't change that, and it doesn't lead to incompatible causal influences on the same entity. The resulting population-environment systems are not the same entity, and the probabilistic causal facts that they give rise to are different. (In particular cases, constraints and modeling complications discussed in chapter 5 may be important to my claim—for example, where there is migration between subpopulations.)

It may seem odd to speak of the "force" of natural selection and the "force" of drift, while saying that they are merely "aspects" of one complex probability distribution. "Force" is often used for entirely distinct causes. However, my view of the relationship between selection and drift fits their mathematical treatment, and it is the mathematical treatment that is at the core of uses of the drift-selection relationship in empirical research and general theory in evolutionary biology (see examples in chapters 2 and 7). The fact that biologists find it useful to deploy the colorful term "force" to describe these interrelated causal influences is not entirely surprising. The correct view is difficult to express in similarly simple, evocative terminology—even for those who clearly understand the mathematical relationship between drift and selection (e.g., Gillespie 2004).

In unpublished work (Abrams Drift MS), I discuss several complications for the view sketched here, including Clatterbuck's (2015) discussion of influences on drift beyond those philosophers had previously acknowledged. The view that I end up with is more nuanced than the one described here, but it is similar.

6.6 Conclusion

I argued in section 6.2 that the conditions that make up an environment can't and shouldn't necessarily be delineated spatially, even though that might be a useful heuristic in some cases. An environment corresponds to a set of probabilities. These are probabilities of outcomes relevant to natural selection being influenced by various possible environmental conditions. In practice, researchers may focus on some subset of those conditions, and they may only have vague ideas about the chances involved. Spatial extents may, of course, play a big role in picking out an environment. This view accords with the pragmatic, jury-rigged nature of scientific research that I highlighted in chapter 5. (Researchers don't need to begin with superhuman knowledge of their subject in order to study it!)

However, researchers delineate environments, though the environmental conditions that matter will also depend on how populations have been delineated. In sections 6.3 and 6.4, I outlined various ways that specifying what counts as a population and its environment can help to pick out different, real causal influences on evolution. Section 6.3 discussed overlapping populations. Section 6.4 discussed different ways of specifying the kind of temporal environmental variation that might be the focus of different research interests. Section 6.5 briefly summarized my view about what drift is, and how it fits into the picture described in this chapter.

As I noted at the beginning of the chapter, statisticalists such as Walsh, Matthen, and Ariew sometimes argue that fitness (and drift) is an artifact of modeling decisions, and as a result don't correspond to causal properties in the world. However, my arguments above (as well as in chapters 5 and 7) provide reasons to think that when models are used in empirical studies, the models help to delineate which aspects of the world are under study, and therefore what causal factors are under study. Researchers don't thereby create the causes. Rather, their research choices determine which existing causes are the focus of their studies. This view also shows why causalists such as Millstein, Otsuka et al., and Gildenhuys should not be troubled by the flexible ways in which researchers define populations and environments in practice. Treating populations and environments as legitimately definable in a wide variety of ways does not imply there are no facts of the matter about natural selection (or drift) as a causal influence on evolution; it just means that these facts will be relative to researchers' choices of questions and research contexts.

Though several arguments in this chapter were based on very simple, abstract models, the point of the arguments was to clarify aspects of practice in evolutionary biology that I've illustrated here or in neighboring chapters. In chapter 7, we'll see, among other things, how biologists use models, statistical inference, measurable token fitnesses, and statistical type fitnesses to learn about causal type fitnesses in an implicitly specified population-environment system.

Fitness Concepts in Measurement and Modeling

7.1 Introduction

In chapter 3, I introduced a five-part distinction between categories of fitness concepts, and I criticized the idea that natural selection can usefully be understood in terms of causal token fitness in the ways that some philosophers have proposed. Chapter 4 gave general principles for thinking about how causal type fitness interacts with environmental variation. In chapter 5, I argued that fruitful empirical practices in evolutionary biology imply that "population" can be used, and is used, in very flexible ways. Chapter 6 explained why this does not lead to conflicts between causal claims about natural selection, despite superficially incompatible ways of specifying populations. The present chapter builds on previous chapters, returning to the fitness categories defined in chapter 3 and focusing on their role in empirical research. I argue we can best understand widespread, typical research practices in evolutionary biology in terms of the claim that

> *causal type fitness* is a property of a trait repeatedly realized in a population-environment system, and it is estimated using *statistical organism-type fitness*, which in turn summarizes *measurable token-organism fitnesses* for organisms with the trait.

This can be divided into the following points:

> *Causal type fitness* is a property of a trait repeatedly realized in a population-environment system.
> Causal type fitness is estimated using *statistical organism-type fitness*.
> Statistical organism-type fitness summarizes *measurable token-organism fitnesses* for organisms with the trait.
> Finally, note that *causal token fitness* usually plays no role in these relationships, and *purely mathematical fitness* can be applied to fitness concepts in various ways.

Despite the variety of fitness concepts used in practice, I believe the pattern I describe above is quite widespread. I begin in section 7.2 with a detailed case study illustrating the pattern. In section 7.3, I discuss potential counterexamples to the thesis: Some studies use hybrids of measurable token and statistical type fitnesses, but they do so in service of a variant of the primary research pattern I describe (§7.3.1). Research on experimental bacterial evolution may seem as if it involves a kind of causal token fitness, but it doesn't (§7.3.2). Some purely theoretical modeling practices sometimes use a kind of causal token fitness concept, but this modeling concept is not what philosophers and others thought was central to understanding evolution (§7.3.3). Finally, I discuss a biological paper that argues that the propensity interpretation of fitness provides the foundation for an area known as life history theory. I argue that this suggestion is mistaken—that life history theory involves another sort of variant on the basic research pattern sketched above (§7.3.4). In section 7.4, I summarize the view developed in this chapter, spelling out some of its implications for recent work in philosophy of biology. Then, after brief concluding remarks (§7.5), the first chapter appendix (§7.6) gives further details about the study described in section 7.2. The second appendix (§7.7) explains why Pigliucci and Kaplan's criticisms of methods like those used in that study don't undermine my arguments.

7.2 Fitnesses in a Study of Contemporary Human Evolution

Byars et al. (2010) applied well-known, widely used methods from quantitative genetics to a human population, arguing that natural selection may be occurring over the course of only a few generations. This study provides a relatively clear illustration of how models and statistical methods are applied in ways that seem to be common in empirical research in evolutionary biology.[1] I introduced the Byars et al. study in section 5.4, discussing the role of population concepts in it, and I'll assume the reader is familiar with that material. Here I'll discuss some of the ways that Byars et al. use fitness concepts. For the sake of clarity, I'll break the study down into what I'll call distinct "steps," but these probably don't correspond to the actual order of tasks during Byars et al.'s research.

7.2.1 BASIC ELEMENTS OF THE STUDY: MEASUREMENT AND STATISTICS

Step 1: Data collection. The first step of Byars et al.'s (2010) study wasn't performed by them: it was the ongoing collection of data on participants in the

Framingham Heart Study (FHS) using surveys, other measurements, and records (also see notes 20 and 21, §5.4.1). Byars et al. only examined data on women in the first three generations of the study, and they found relationships between (what I would argue is) causal type fitness and measurements of women's total cholesterol, blood pressure, and blood sugar, among other characteristics. However, I'll focus on relationships to age at first birth—that is, the age at which a woman had her first child—because the relationships that Byars et al. found with this property were clearest.

Step 2: Heritability. In the second step, Byars et al. modeled heritability relationships according to variables available in the FHS data, using the data to calculate heritability relationships between traits that they studied.[2] This showed that age at first birth had a relatively high heritability, in that among the individuals studied, there was a high correlation of age at first birth between parents and offspring. Age at first birth also showed correlations with other traits, and thus could have been influenced by factors that influenced these other traits. (Byars et al. didn't assume that the heritability relationships were entirely due to genetics; they allowed that these relationships could have been at least partly mediated by cultural influences of parents on their children. See note 13, §0.3.2.)

Step 3: Record LRS. The third step of Byars et al.'s study was to record the lifetime reproductive success (LRS) for women in the first generations of the FHS study. Byars et al. measured LRS as the number of children recorded for each woman at the time of menopause. LRS is also sometimes called fitness, fertility, fecundity, or realized fitness, among other things. It is clearly a *measurable token fitness* concept, since it is a fitness property that can be readily measured for a token individual.

Step 4: Regression. The fourth step was to model LRS as a function of trait values. Among other things, Byars et al. calculated a partial regression coefficient from a multiple linear regression of LRS on age at first birth, controlling for various other factors. This partial regression coefficient is the slope of a line that best fits, on average, the relationship between age at first birth and LRS in the data. It is an estimate of what you might think of as a functional relationship. Byars et al. found that among the women they studied, on average the younger a woman was when her first child was born, the more children she had. (The value of the partial regression coefficient was −1.267, indicating that an increase of one year in a mother's age when she had her first child corresponded, on average, to 1.267 fewer children.)

Note that the partial regression coefficient is a mathematical quantity calculated directly from the LRS values for all of the women in the study, giving a statistical average relationship between a phenotypic property of women in the study and a measurable token fitness, LRS. The regression coefficient of LRS on age at first birth is thus a *statistical type fitness* for this trait. It is a population-wide summary of the relationship between measured ages at first birth and measured lifetime reproductive successes. (For further details see the first appendix to this chapter, §7.6.)

7.2.2 SELECTION GRADIENTS AND PREDICTIONS

Note that Byars et al. were using the FHS data in a particular model of relationships between (only) some traits that had been measured and a particular measurable token fitness property, LRS. I summarize further aspects of this model and its use next. I want to emphasize, though, that like most models in evolutionary biology (§2.1), this one idealizes from reality in that it ignores many complexities in the population-environment system under study.

Step 5: Selection gradient. What I'll call the fifth step of Byars et al.'s study is their treatment of the regression coefficient as an *estimate* of a *selection gradient*, a measure of the extent to which a possible individual with a trait varying along a continuum—age at first birth, in this case—would be *likely* to have more or fewer offspring:

> Selection gradients measure the extent to which individuals with a given value of a quantitative trait *tend* to have higher or lower fitness (LRS). (Byars et al. 2010, 1788; emphasis added).

Note that "fitness" here is LRS; that is, it is a measurable token fitness. The regression coefficient, calculated directly from the data, is treated as an estimate of something else that can't be directly measured:

> The direct effects of selection are captured by the linear selection gradients (β) estimated as partial regression coefficients. (Byars et al. 2010, 1788).[3]

The selection gradient doesn't measure a relationship between actual token parents and their offspring, per se, since it is merely estimated from statistics concerning patterns of association between them.[4] The selection gradient is a property of the population in a deeper sense.[5] Specifically, a selection gradient is the probability-weighted average change in a trait due to fitness differences among possible variants of that trait (excluding effects of selection on other traits that can influence the first trait indirectly[6]). If we understand the probability involved here as causal probability, then the selection gradient is a *causal type fitness*: it is a causal property of heritable traits (ages at first birth) that reflects probabilistic influences of the trait on changes in its future distribution in the population (in this case via influence on LRS). Recall that on my view, traits are potentially realized properties of a population-environment system. So, on this view, selection gradients are defined by an underlying causal probability distribution, realized by the population-environment system considered as a chance setup. Selection gradients summarize probabilities of (children's) trait realizations conditional on earlier (parent's) trait realizations. I'll provide additional motivations for this view below.

Step 6: Prediction. The sixth step performed by Byars et al. was to use the estimated selection gradients and other information to predict changes in the FHS population in future generations. In the case of age at first birth, Byars et al. predicted that in ten generations, the average age at first birth for women in the population would probably be reduced from 26.18 years to roughly 25.74 years, with a 0.95 chance that this average would lie between 25.26 and 26.22 years. (Note that this prediction reflects not only the selection gradient for age at first birth but also selection gradients for other traits, as well as other factors that might indirectly influence age at first birth.)

The fact that Byars et al. gave a prediction is relevant here because it illustrates how the idea of natural selection acting as a cause is implicit in empirical practice. Byars et al. wrote:

> We measured significant linear selection gradients acting to reduce . . . age at first birth. . . . These results strongly suggest that *natural selection is acting on the women of the FHS* through differences in the number of children they had during their reproductive ages. (Byars et al. 2010, 1788; emphasis added)

Byars et al. claimed that natural selection was "acting" on the population—because of the selection gradient they estimated. First note that "age at first birth" refers to the simple average of age at first birth for all women in the population at a given time (e.g., after ten generations). Though this is just a simple average, the prediction is based on the assumption that different distributions of traits of women in the population ten generations into the future have different probabilities. The probabilities are summarized by the selection gradients, and on my view, we should take these probabilities to be generated by the overall character of the population-environment system that is reflected in the FHS data.[7] By calculating predicted values for traits such as age at first birth from selection gradients (and other information), Byars et al. implicitly assumed that manipulations to selection gradients could effect changes in future trait distributions. Since selection gradients are defined in terms of probabilities of changes, manipulating a selection gradient implicitly involves manipulating probabilities. Thus the use of selection gradients

to predict future changes in the population implies that manipulating probabilities can manipulate future distributions of traits. These probabilities are
therefore causal probabilities (§2.2).[8] That is, this research is based on assumptions that imply that the system studied seems to realize causal probabilities.[9]

Apart from Byars et al.'s relatively unusual treatment of populations
(chapter 5)—including the fact that one can understand their research as a
study of an entire population rather than a sample—their study is method-
ologically similar to many other empirical studies in evolutionary biology.
First, applications of such quantitative genetic methods involving estimation
of selection gradients and predictions for empirical populations are common
in evolutionary biology.[10] (See the second appendix to this chapter, §7.7, for
discussion of a challenge to such methods.) These methods are also widely
used for animal and plant breeding (Falconer and Mackay 1996; Lynch and
Walsh 1998), in which controlled artificial selection produces concrete results. Second, many—probably, nearly all—empirical studies of natural selection using population genetic rather than quantitative genetic methods can
also be understood in terms the three-stage use of fitness concepts I have
described. Most of the empirical studies mentioned in this book follow exactly this pattern, and further illustrations can readily be found in journals
that feature empirical research in evolutionary biology. Thus, given the success of research in evolutionary biology in general, we have evidence (§2.4)
that causal probabilities are typically realized in the population-environment
systems studied in evolutionary biology.[11]

The argument for causal probability in population-environment systems
in section 2.3 inferred the existence of causal probabilities from close parallels
between roles of probabilities in simulations and population-environment
systems modeled by the simulations. The preceding argument is different. It
infers the existence of causal probabilities from practices of statistical inference. In the following sections, I extend and deepen this conclusion about
causal probabilities and empirical research in evolutionary biology.

7.2.3 INTERPRETING P-VALUES

In addition to the explicit use of "estimate" in the text, the idea that the selection
gradient itself—that which is estimated—is not directly measured is also supported by the way that each estimate is qualified with a p-value (see §2.4.2). The
p-value in the case of age at first birth is the probability that a similar statistical
pattern (a partial regression coefficient) observed between age at first birth and
LRS would have been observed in the data just by chance—that is, even if there

were no causal relationship between age at first birth and LRS (e.g., Bulmer [1967] 1979; Cox 2006).[12] A small *p*-value means that this probability is low, and evolutionary biologists generally interpret a low *p*-value as a reason to accept that the estimate should be treated as approximately correct. The idea is that if there was in fact no reason that changing the age at first birth would have made a woman more likely to have a more children, then it still could happen, just by chance, that those women with lower ages at first birth ended up having more offspring. In theory, this could be the source of what Byars et al. observed in the FHS data. The fact that the *p*-value that they found was low—less than 0.0001—means that this outcome would be very improbable if there was not in fact influence of age at first birth on LRS.

We can understand the *p*-values in Byars et al.'s study as probabilities generated by a hypothetical underlying population-environment system—one that would have existed had there been no causal influence of age at first birth on LRS (see Cox 2006; Mayo 2018). On my view, *p*-values referenced in empirical studies of evolution are causal probabilities, which would be realized by possible population-environment systems in which certain chancy causal influences were absent. Biologists in effect treat small *p*-values as evidence that the actual population-environment system is one in which such a causal influence *is* present.

7.2.4 FITNESSES AND PROBABILITY IN EMPIRICAL RESEARCH

To summarize, Byars et al. (2010) treated LRS as a function of women's trait values as recorded in data provided by the Framingham Heart Study. The LRS value for each woman is a *measurable token fitness*. From LRS values and trait values, Byars et al. calculated regression coefficients relating various traits to LRS—including age at first birth. The regression coefficients are *statistical type fitnesses* of the traits. They measure relationships between each trait and LRS values. Finally, Byars et al. treated the regression coefficients for LRS on each trait as estimates of something that could not be directly observed: the selection gradient for each trait. These selection gradients are *causal type fitnesses*; they are constituted partly by causal probabilities realized by the underlying population-environment system. Thus the example illustrates the relationships stated at the beginning of the chapter:

> Causal type fitness (e.g., a selection gradient) is a property of a trait (e.g., age at first birth) repeatedly realized in a population-environment system.

It is estimated using *statistical organism-type fitness* (e.g., a regression coefficient of LRS on age at first birth).

Statistical organism-type fitness in turn summarizes *measurable token-organism fitnesses* for organisms with the trait.

Causal token fitness usually plays no role in this relationship.

It's important to acknowledge that unlike Byars et al's study, most empirical studies in evolutionary biology perform statistics on a sample of individuals from a population rather than on an entire population (§5.4.1). Sampling introduces a layer of probability that is plausibly not present in Byars et al's study. For example, in the Samuk, Xue, and Rennision's (2018) study (chapter 5) of effects of predation on threespine stickleback fish, populations were defined by ponds stocked by researchers, but the fish that were measured were sampled from each pond. Although I won't argue for the point here, on my view, when researchers sample individuals from a population, they usually choose the individuals according to causal probabilities of one kind or another.[13] However, these are usually different from probabilities in the underlying processes that are studied. Sampling probabilities in turn implicitly determine probabilities that the statistical type fitnesses of the sample are likely to be similar to statistical type fitnesses measured on the whole population. Thus, in most empirical studies in evolutionary biology, estimation of causal type fitnesses from statistical type fitnesses depends both on sampling probabilities and on population-environment probabilities.[14] The Byars et al. example is useful because it illustrates the point that there is estimation of underlying fitnesses—causal type fitnesses, I claim—even when sampling probabilities play no role in justifying that inference.[15]

Note that there are ways of getting evidence of natural selection without referencing fitness at all. One can use F_{ST} for this purpose (§1.1), and the Santos et al. (2016a) study described in chapter 2 provides another kind of example. There are recently developed methods in which complex patterns in genetic data provide evidence of natural selection. I discussed one such method in Abrams (2015), but there are many others (e.g., Li and Stephens 2003; Liu et al. 2013; Lynch and Walsh 2018). Endler (1986, chap. 3) discusses a variety of older methods. It seems that all such methods for deriving evidence for natural selection are nevertheless like the methods described above, in that they depend on measuring and summarizing properties of actual token organisms in one way or another and then estimating underlying facts about natural selection. My claim here, though, is only that when fitness concepts *are* used, they are usually used in the way I've described.

7.3 Potential Counterexamples

Next, I discuss several apparent counterexamples to claims I've made about the roles of fitness concepts in evolutionary biology.

7.3.1 HYBRID AND MISSING FITNESS CONCEPTS

In some cases, biologists measure fitness using concepts that are hybrids of measurable token and statistical type fitness concepts. In chapter 6, I discussed some aspects of Hamann, Weis, and Franks's (2018) study of field mustard plants. The authors described the mass (not the count) of seeds per plant as "a component of fitness" (2687).[16] Specifically, they calculated "relative fitness" (2686) in the following manner: First, the researchers calculated the average mass of seeds produced by two plants that were each grown from seeds deriving from a single maternal parent plant. (Note that there are three generations involved in this measurement: a parent, two offspring, and the offspring's seeds.) The flowers of the maternal parent were probably fertilized by pollen from different paternal parents, so the average seed mass only partially represented effects of the maternal plant's genes. Hamann, Weis, and Franks then calculated relative fitness by dividing this two-individual average seed mass value by the average seed mass for all plants in the same treatment group (defined by source population, year of seed collection, and amount of water provided). This notion of relative fitness is a hybrid of a measurable token fitness concept—since it can be viewed as a measurable fitness of the maternal parent plant—and a statistical type fitness concept—since it's based on an average between two individuals' reproductive output. This is not really problematic from my point of view, though, since I see the primary point of measurable token fitness concepts in biological practice as providing a basis for calculating statistical type fitnesses, which are then used to estimate causal type fitnesses—as Hamann, Weis, and Franks did later in their paper, using methods similar to those described in section 7.2.2. That is, it's not essential to my approach that there always be a clean distinction between measurable token fitness and statistical type fitness concepts. It is important, I believe, to distinguish measurable token fitness or statistical type fitness concepts from causal token fitness concepts in order to avoid misunderstandings such as those mentioned below and in chapter 3. In particular, the distinction between measurable token fitness and causal token fitness helps to prevent an unjustified slide from a kind of fitness that directly depends on measurement of simple property of an individual to a kind of individual fitness that is causal. (Pence and Ramsey's [2015] criticism of Sober [2013], which I describe below in §7.4, illustrates this kind of mistake.)

7.3.2 FITNESS IN EXPERIMENTAL BACTERIAL EVOLUTION

There is a significant body of research involving experimental evolution of microorganisms. Richard Lenski's "Long-Term Experimental Evolution" (LTEE) project, which studies populations of *E. coli* evolving over tens of thousands of generations, is widely known (e.g., Lenski et al. 1991; T. F. Cooper, Rozen, and Lenski 2003; Wiser, Ribeck, and Lenski 2013; Wiser and Lenski 2015; Tenaillon et al. 2016; Peng et al. 2018). This kind of research enables some interesting methods, such as freezing a sample from a population and then comparing its growth with the same population at a later time, or "replaying" evolution from a specified point at which members of a population have been frozen. There are quite a few other projects that use microbes in experimental evolution (Lenski 2017). The LTEE and other microbial evolution projects have generated a large number of papers, and I can't do justice to this area's potential philosophical interest (e.g., O'Malley and Parke 2018).

The LTEE research measures evolution in multiple populations of *E. coli*. Each population is grown from clonal offspring of an initial founder bacterium. From these bacteria, Lenski et al. (1991) isolated two bacteria that differed only by a single neutral mutation—that is, a mutation that should make no difference in fitness in the test environments.[17] (These bacteria were unable to undergo horizontal or lateral gene transfer—i.e., the transfer of genes from one bacterium to another.[18]) The neutral genetic difference thus allowed the researchers to distinguish bacteria of the two kinds even when grown in the same culture.[19] Thus the researchers could, when desired, subject genetically distinct lineages to exactly the same environment by putting them together in a common culture. In this case, given that the bacteria didn't initially differ in any way likely to affect potential for growth and adaptation, differences between the growth rates of the two populations were thought to depend primarily on differences in mutations acquired over time.

The LTEE project seems as if it might provide an interesting case for thinking about fitness. First, each population consists of lineages that have evolved from a single bacterium. Ramsey and Pence have proposed that the key to understanding long-term evolution would be to define causal token fitness in terms of many generations of descendants of an initial token organism (Ramsey 2006; Pence and Ramsey 2013; see Thoday [1953] and W. S. Cooper [1984] for earlier long-term fitness concepts). Thus one might consider fitness measurements of LTEE bacteria to be measurements of causal token fitness for a founder bacterium. Second, Lenski et al. (1991) grew multiple "replicate populations" from each founder bacterium, so there are multiple versions of what might be viewed as "the same" lineage, in the sense that those lineages each evolved from genetically identical bacteria (cf. §1.1). Thus one might view the distinct populations as providing samples from a probability distribution over possible growth rates for descendants of the initial parent bacterium. That is, measurable token fitnesses based on the populations that descended from a single founder bacterium could be used as an estimate of the underlying causal token fitness of that ancestral bacterium (e.g., by averaging the growth rates).

However, the fitness measures used in the LTEE project are not, fundamentally, long-term fitness measures. The LTEE uses fitness measures defined in terms of short-term growth rates of populations throughout—as reported both in the original paper (Lenski et al. 1991) and in Lenski's (2017) review. Lenski et al. (1991) used several fitness measures, but all were defined in terms of growth rates measured by checking the size of a population at two points in time (e.g., a day apart). Then the authors compared the growth rates of two populations to generate a relative fitness measure. Wiser and Lenski (2015) described additional methods that are useful for comparing the fitnesses of populations that have been evolving for a long time; these methods are a bit more involved but still depend on measuring changes in the sizes of populations over relatively short periods. It's worth noting that because of the ability to freeze and revive samples taken from a population at an earlier time, Lenski et al. (1991) were able to compare the fitnesses of the same population at different times, using measurements of the kind just mentioned. But that sort of comparison doesn't measure the probable long-term growth of the initial population or its bacterial progenitor; it just compares short-term fitness measurements at two times.

Moreover, Lenski and his collaborators routinely treated the fitnesses they measured as fitnesses of types not fitnesses of token organisms. For example, Elena and Lenski wrote:

> If they are measured with sufficient temporal resolution, then fitness and traits that are correlated with fitness (such as cell size in *E. coli*) change with a step-like dynamic. Each step probably corresponds to the spread of a beneficial mutation. (Elena and Lenski 2003, 460)

Here a (beneficial) mutation (i.e., a genetic variant) is clearly an inheritable type, since it can "spread." Each stepwise change in fitness "probably corresponds to" such a mutation—that is, the fitness is influenced by the mutation, a genotype. Elena and Lenski also explicitly described fitness as "the average reproductive success of a *genotype* in a particular environment" (Elena and Lenski 2003, 458; emphasis added). So the growth rate measures that Lenski and collaborators used are measures of type fitnesses. These growth rates are, in fact, primarily ways of *measuring* fitnesses—that is, they are statistical type fitnesses. This is a case where statistical type fitnesses can be measured directly, without first measuring measurable token fitnesses (see §7.3.1). Lenski and his collaborators also seemed to treat statistical type fitnesses as estimates of what I take to be a causal type fitness; this is suggested by Lenski et al.'s (2015) use of p-values to determine whether increases of measured fitness in the twelve populations could have been due to chance (cf. §7.2.3).[20] Thus, despite the fact that the LTEE uses replicate lineages to study evolution, they are still using statistical type fitnesses, apparently to estimate causal type fitnesses. Causal token fitnesses are not part of this project.

If the LTEE researchers were really interested in probabilistic fitness relationships between lineage progenitors as such—that is, if they were interested in causal token fitnesses—then it would have made more sense to study lineages from progenitors that differed in ways that could matter to fitness. Even so, given the number of fitness-affecting mutations that occur on the time scales in LTEE studies (two thousand generations, sixty thousand generations, etc.), what would be studied would depend heavily on mutations rather than initial types alone. LTEE researchers treat different lineages as realizing different types because the lineages have accumulated different mutations, giving them different fitnesses: "Although founded by the same clone, and evolving in identical environments, replicate populations often diverge from one another in their relative fitness" (Elena and Lenski 2003, 461). This claim is most easily understood as a claim about statistical type fitnesses and causal type fitnesses estimated by them.

Finally, even if one were to design an LTEE-style experiment designed to use replicate lineages to measure fitness differences between a few token progenitor organisms, it's not entirely clear what value it would have for evolutionary biology. Arguments in chapter 3 imply that given most real-world environmental variation and most real-world genetic variation, studying a few causal token fitness differences wouldn't tell us much about evolution. The magnitude of differences between causal token fitnesses depends more on variation in circumstances, and this kind of fitness difference is not very inheritable. What's interesting and informative are fitness differences between types, and that is what the LTEE seems to study among other things (Lenski 2017).

7.3.3 THE CASE OF THE PRICE EQUATION

I've argued that causal token fitnesses seem to play no role in empirical research in evolutionary biology, but perhaps they have a valuable role in more abstract modeling domains. The first case I'll consider illustrates a way in which causal token fitnesses might play a role in modeling—but I'll argue that we have to be careful not to misinterpret this role. Then, in the next section, I'll discuss a case that illustrates what I see as a misunderstanding about causal token fitnesses in modeling.

I noted in section 3.2.2 that the same terms in the Price equation can be (and have been) described as representing traits and individuals. I illustrated the point in section 3.6 using a simple version of the Price equation (3.1), which I reproduce below. This equation is closely related to the Lande-Arnold framework (Lande and Arnold 1983; Rice 2004) that is the basis of Byars et al.'s (2010) model.

$$\Delta \bar{z} = \text{cov}(z, w) = E_i[(z_i - \bar{z})(w_i - \bar{w})].$$

I explained in section 3.6 that z_i could either be interpreted as representing the ith competing trait or as the trait of the ith individual, which might be the same trait as that of some other individuals but need not be. The fitness terms w_i have the same ambiguity, representing either a trait fitness or the fitness of a token organism. Thus, without further interpretation, the fitnesses in the Price equation are purely mathematical fitnesses (§3.2.2).

Price's (1970, 1972) own discussion of his (more complex) equation was in terms of fitnesses of individuals—that is, of token organisms. Price treated fitness as the number of gametes that an individual contributes to the next generation. Thus he allowed organisms with the same trait to have different numbers of offspring, and therefore different fitnesses w_i. These fitnesses are thus token fitnesses of some kind. In terms of my categories, Price's fitnesses can be viewed either as causal token fitnesses or as measurable token fitnesses. I see no reason to think that Price was even implicitly aware of such a distinction. However, to the extent that Price or others—possibly Rice (2004)—treat Price equations as if fitnesses were causal properties of token individuals, causal token fitness would have at least a vague role in theoretical work in evolutionary biology. Moreover, one might argue that whatever the limitations of empirical research, modeling is sometimes directed at elucidating the underlying character of evolutionary processes. Thus uses of models in which causal token fitnesses are assumed could be viewed as evidence that biologists take evolutionary processes to involve causal token fitnesses.

However, note that the conception of token fitness involved here is very different from the kind of causal token fitness that philosophers of biology seem to attribute to token organisms. Neither Price (1970, 1972) nor Rice (2004) assigned to token individuals probability distributions over possible numbers of offspring. In fact, since both authors identified token fitnesses with numbers of offspring, the fitnesses look like a modeler's version of measurable token fitnesses. Alternatively, by defining fitnesses in terms of numbers of offspring, Price and Rice may simply have been using frequencies as a way of modeling probability. This is fine as modeling, up to a point, but it's not how empirical researchers usually proceed, as we've seen. And as a basis for causal type fitness, it would be subject to problems about fluctuating fitnesses described in section 3.3.4. It is nevertheless a useful conceptual heuristic to treat traits and fitnesses in the Price equation as attached to token individuals, and to treat probabilities as frequencies: it can be easier to think about probabilities as if they were frequencies. There's nothing wrong with a model that uses a simplified representation for the sake of developing understanding. However, if the Price equation is intended to capture causal relationships, the probabilities must be understood as causal probabilities, which are more than frequencies. The token-organism plus frequencies interpretation of the Price equation should be taken literally only if the goal is to use the Price equation to summarize what has actually happened. Similar remarks should apply to many other models that model fitnesses as properties that vary between individuals of the same type.

7.3.4 LIFE HISTORY THEORY MODELS

A paper by two biologists, McGraw and Caswell (1996), seems to advocate explicitly for the relevance of the propensity interpretation of fitness (§0.3.2) to certain models used in empirical research.[21] McGraw and Caswell cited several philosophical works on the propensity interpretation of fitness (PIF) (Brandon 1978, 1990; Mills and Beatty 1979; Burian 1983; Brandon and Beatty 1984; Sober 1984). They also cited works on the propensity interpretation of probability by Popper (1959, 1983, 1990), whose ideas about probability are important to the PIF. Since McGraw and Caswell's paper seems to make claims that directly contradict several of my conclusions, it's worth discussing in detail. It will require some unpacking to explain why I think the paper involves confusions, which I take to be understandable at the time.

McGraw and Caswell's paper falls within what is known as life history theory, which studies influences on evolution of organism states—properties and behaviors—that can take place at different times in organisms' lives (Stearns

1992; Caswell 2001). I want to begin by clarifying how empirical research in life history theory fits into the framework that I've developed. In life history research, it's important to track various properties of a single individual over time. That requires a concept of an individual over time, and measurable token fitnesses and other relevant measurements have to be made *on the same token organism* at different points in its life. General properties of relationships between these actual patterns can then be derived from the measurements of token individuals, and these general statistical properties can then be used to estimate probabilities for various sequences of property realizations by individuals in a population-environment system, as in section 4.2.3. So the measurements of different states at times have to be done in such a way as to track different individuals. However, that doesn't mean that the ultimate conclusions of such a study are about token organisms per se. A series of life history events at particular times or in particular sequences is a type; it is something that can be realized multiple times in the same population-environment system.

McGraw and Caswell described methods that seem to treat changes in the frequencies of organism's properties as dependent on modeled individual organisms. Consider the following quotation. (It's not necessary to understand all of the mathematical terms.)

> The [dominant] eigenvalue $\lambda^{(m)}$ is a logical measure of fitness because it is an *estimate* of the asymptotic growth rate of a collection of individuals with the propensities to survive and reproduce of individual m; that is, it is an estimate of the propensity fitness of individual m. The value $\lambda^{(m)}$ is the population growth rate of the individual. (McGraw and Caswell 1996, 50; emphasis in original)

Note that McGraw and Caswell described a model of "a collection of individuals," each of which has the same "propensities to survive and reproduce" of a *single* individual m. Thus, this is a population in which every member has the same fitness, which is at first glance a causal token fitness. A reasonable reading here is that "population" refers to a subpopulation of individuals whose fitnesses are identical to that of m.

When McGraw and Caswell mentioned the dominant eigenvalue, they were alluding the fact that their models were matrix models that tracked numbers of individuals in particular states at times during an organism's life. For example, a transition matrix for a model of a plant population might represent probabilities of plants shifting, within a certain period of time, from a seed state to a preflowering mature state, or from the latter to a flowering state, and so on (see Caswell 2001). These models can be used to calculate numbers (or probable numbers) of individuals in states at subsequent times.

The dominant eigenvalue represents the eventual average growth rate of a population as modeled by such a matrix. This matrix isn't, generally, a perfect representation of what happens in a real population, though, which is one reason the growth rate indicated by the dominant eigenvalue is, as McGraw and Caswell said, an estimate. This growth rate, context makes clear, is a population-level fitness measure, derived by modeling a population (or subpopulation) of organisms as if they all had the (causal token) fitness of a single individual. Since every member of the modeled population has the same fitness, the leading eigenvalue is thus an estimate of individual m's fitness: "that is, it is an estimate of the propensity fitness of individual m." So it is this individual's fitness that can be understood in terms of the PIF, according to McGraw and Caswell.

I think that McGraw and Caswell's invocation of "propensity fitness" either involves a misreading of classic works on the PIF or involves some confusion about what is being modeled. In real populations, there won't be multiple individuals with the same causal token fitness (§3.3), and advocates of the PIF generally have not claimed that there would be. I argued in section 3.4 that this is so even for clones. So McGraw and Caswell are trying to claim that a fitness concept for an entire class of organisms—growth rate—is based on the assumption that every organism in the subpopulation has the same causal token fitness. Notice that since the growth rate in this model is a fitness defined as an average over multiple, identical individuals, it is a type fitness. So it looks like McGraw and Caswell are defining a kind of causal type fitness in terms of causal token fitness. This seems reasonable in the context of an abstract model, but such subpopulations don't exist in reality.

McGraw and Caswell also talked about using matrix models of the kind they described in empirical studies:

> This measure of fitness corresponds to the way that fitness is measured in more traditional ways by statistical measurements on collections of individuals. In such studies, a set of individuals that are considered to be experiencing the same propensities is defined. (McGraw and Caswell 1996, 51)[22]

"This measure of fitness" in the first sentence refers to what McGraw and Caswell called "relative individual fitness," which is defined in terms of the growth rate. So this, too, depends on what ought to be viewed as a fitness of types. The authors claimed that this "corresponds to" (i.e., models, or can be defined from) what I view as a statistical type fitness ("statistical measurements on collections of individuals"), where "a set of individuals that are considered to be experiencing the same propensities is defined." This seems to be a claim that when researchers are studying an empirical population, they treat individuals as all having the same probabilities of outcomes. That seems somewhat unlikely to me, but it may be that some researchers do conceive of similar organisms as all having the same probabilities of outcomes. The idea might be that individuals that share certain traits have the same probabilities of transitioning between states. However, in that case these should *not* be viewed as probabilities relative to each token organism and its circumstances. My arguments show that such measurements on individuals provide information on their common traits—not on their causal token fitnesses.

In fact, later in the same paragraph McGraw and Caswell seemed to acknowledge that measurements in empirical studies are of individuals with different properties:

> More usually, the individuals consist of a sample of genotypes obtained from a population of interest, and the resulting measure of fitness is a mean over the genotype distribution. In any case, the fates of those individuals are followed over time, and the vital rates (i.e., propensities) estimated by maximum-likelihood methods. (McGraw and Caswell 1996, 51)

In the first sentence, McGraw and Caswell mentioned the "mean over the genotype distribution," which I take to be the mean value of an outcome for actual individuals with different genotypes. In the second sentence, it's possible that McGraw and Caswell were claiming that researchers estimate probabilities of outcomes for token organisms using statistical (maximum-likelihood) methods. However, I don't see how that is possible, except by using information about many individuals to impute properties to a single individual. I think it's more reasonable that what they are describing is a causal type fitness estimated from data about multiple individuals.

Note that McGraw and Caswell also said that

> using clones, it may be possible literally to replicate individuals and thereby estimate probabilities for $P_i^{(m)}$ or means for $R_i^{(m)}$ to obtain estimates of fitness that more closely correspond to the theoretical (propensity) fitness of the individual. Indeed, theoretical life history studies have defined fitness as the population growth rate of a clone (see, e.g., Stearns and Crandall 1981). (McGraw and Caswell 1996, 51)

If causal token fitness is the kind of fitness that's intended here, the claim that one can estimate reproductive propensities of clones from a series of observations is reasonable, though on my view incorrect (§3.4). The paper by Stearns and Crandall cited at the end of the passage developed mathematical models:

> To keep the models tractable, we assumed that we could deal with a population as though it were a set of asexually reproducing, haploid clones, with

each clone empowered with a different life-history [i.e., with different states at times]. (Stearns and Crandall 1981, 455)

Stearns and Crandall clearly intended the assumption that members of the population are clones as an idealization, not as a description of reality. Nevertheless, the fact that Stearns and Crandall considered it reasonable that clones could produce different outcomes might indicate that they think token organisms have probability distributions over outcomes regardless of environmental circumstances. However, it's also compatible with my view that such variation in outcomes is probably due to circumstance variation. Coming back to McGraw and Caswell, suppose that they do think that one could estimate probabilities for outcomes not dependent on *circumstances* (§3.3) by measuring outcomes of clones. That would just show that earlier passages in which McGraw and Caswell seemed to say that causal token fitnesses can be measured or estimated are at best misleading. Most organisms are not clones, so it's unlikely that they share enough genetic and other material—let alone circumstances—to have the same causal token fitness.

None of this is meant to imply McGraw and Caswell do not know what they are doing in population modeling and its empirical applications in practice. What I am suggesting is that the relevant practices, about which McGraw and Caswell are certainly well informed, use models in ways that are at odds with philosophical and theoretical claims that the authors made in this article. Moreover, I suspect that the fact that life history traits have to be measured on individuals that are tracked over time makes it easier to confuse the idea of fitness as a property of a trait with fitness as a property of an individual. Without having available the kind of distinction between fitness concepts that I introduce, and with only the PIF tradition as a potential philosophical foundation for evolutionary biology, it's understandable that McGraw and Caswell would suggest that causal token fitness is implicit in life history research.

7.4 Implications

In this chapter, I argued that empirical research in evolutionary biology that uses fitness typically involves inferring approximate values for *causal type fitnesses* from *statistical type fitnesses*, which summarize *measurable token fitnesses* (§7.2), and it has no general role for causal token fitness. I used a case study introduced in chapter 5 to illustrate this point, but it's easy to find many other examples of the same pattern. If, as I claim, this pattern of inference and some close variants of it (see §7.3.1) are pervasive in evolutionary biology, that fact gives us a reason to think that the pattern illustrates a generally correct way of understanding relationships between fitnesses, natural selection, and evolution in the world: otherwise practices using this pattern would probably have been abandoned (§2.4). Note that while the example I gave uses models from an area known as quantitative genetics, the same points could have been made using models from population genetics. Other examples in this book, some of which I didn't discuss in detail, could easily illustrate this point.[23] I also maintained that my argument provides additional reasons to think population-environment systems realize causal probabilities. In section 7.3, I discussed several apparent counterexamples to the claim that fitness concepts are used in the way that I claim they are, but I argued that in each case that the supposed counterexample either supports one of the "close variants" or is not a counterexample for other reasons.

I think my view clarifies the nature of evolutionary biology and of evolution, and I hope it will be embraced by some biologists. However, the most direct implications of the view concern philosophy of biology. While I can't discuss all ideas or works to which my view might be relevant, a few comments on some recent papers should help to clarify some implications of my view.

- Sober (2020) remarked that "traits cause survival and reproductive success only in virtue of their attaching to objects"—that is, to actual organisms (Sober 2020, 9). Sober is right, in the sense that only actual property realizations can cause actual effects. However, understanding the causes of *evolution* as *causes* requires understanding more than what happens to occur: probabilities of possible instances of traits realized in other circumstances matter as well. On my view, traits are realized by a population-environment system itself—that is, by the aspects of a population-environment system that we call "organisms." The advantage of this view is that a population-environment system is a chance setup with complex outcomes, realizing a variety of properties relevant to evolution. A population-environment system realizes traits as traits of actual organisms—*but also* determines probabilities for a variety of the *possible* realizations of the same (and other) traits. Actual organisms, on the other hand, cannot by themselves realize probabilities of possible outcomes in which those organisms don't exist, or in which they would encounter different circumstances. Causal probabilities of those other conditions are crucial to understanding causal organism-type fitness. It is these causal probabilities, which are embodied by the population-environment system, that provide what is missing from any account of natural selection that focuses only on actual organisms.[24] My view nevertheless allows properties based on causal probabilities (e.g., causal type fitnesses) to be estimated from what actually happened (using statistical inference).

- Many of Sober's (2020) conclusions are consistent with mine, but he initially seems to endorse Mills and Beatty's (1979) thesis that trait fitnesses are equal to averages of (causal) fitnesses of actual token organisms (Sober 2020, 4). Later Sober writes that "each of those token organisms has just one lifetime, but averaging over numerous lifetimes can provide a meaningful estimate of the fitness of the trait those organisms share" (Sober 2020, 5). That the average is used as an "estimate" suggests that trait fitness is more than a property defined by a collection of actual organisms, however. Sober's (2020) assertions on pages 4 and 5 thus seem to imply that the average is both trait fitness and something other than trait fitness. On my view, we can preserve what is right in each claim: an average of actual *measurable* token fitnesses is one way to define statistical type fitness, which can be used to estimate causal type fitness (i.e., trait fitness in a causal sense).

- Pence and Ramsey (2015) argued that Sober (2013) was wrong to claim that token fitness does not play a role in evolutionary biology. One of their arguments was that biologists do measure token fitnesses, as illustrated by examples in Endler (1986). However, examination of Endler's book or Pence and Ramsey's discussion makes it clear that Endler's examples of measurement involved measurable token fitness, while Sober seemed to be denying a role to causal token fitness.

- Because statisticalists argue that natural selection is not a cause of evolution (e.g., Walsh, Ariew, and Matthen 2017), they argue that there are no causal explanations of evolution in terms of natural selection (or drift etc.). Statisticalists sometimes argue, though, that there are mathematical or statistical—not causal—explanations of evolutionary patterns in terms of properties of populations (Matthen and Ariew 2002, 2009; Walsh 2007; Ariew, Rice, and Rohwer 2015; Walsh, Ariew, and Matthen 2017). They may be right that there are mathematical explanations that connect empirical premises to empirical conclusions (see note 28, §0.3.6). However, if I'm right that arguments in this chapter provide good reasons to think that causal type fitnesses exist, it's reasonable to think that fitness differences between traits, and hence natural selection, are legitimate parts of causal explanations of evolution.

- Statisticalists also sometimes claim that legitimate generalizations concerning properties of populations can be found only in what they call "the modern genetical theory of natural selection" (Ariew, Rice, and Rohwer 2015, 641), "Modern Synthesis" models of natural selection (Walsh, Ariew, and Matthen 2017), or "population genetics" (Matthen and Ariew 2002; Matthen 2009), which is "the mathematical form of the theory of natural selection with the additional assumption that inheritance is Mendelian" (Matthen 2009, 477). The arguments and examples I've given should make it clear that the mathematical parts of evolutionary biology are intimately involved in empirical research and, in particular, in research that infers causal facts about populations. "Population genetics" (along with quantitative genetics) doesn't constitute an isolated domain. It is the skeleton that allows the body—the empirical study of microevolution—to function as it does.

7.5 Conclusion

So far I have provided what I believe is a consistent, unified conception of evolution as taking place in real but pragmatically delineated population-environment systems. I've argued that these realize the causal probabilities in terms of which natural selection and fitness can be defined. I claim that this picture, including the existence of such causal probabilities, is supported by its coherence and consistency with evolutionary theory and successful empirical practice. I have, however, not given an account of the nature of the causal probabilities nor what they depend on. It would be valuable (though not essential) to do so. The next two chapters explore some possible responses to this challenge.

7.6 Appendix: Details of Byars et al.'s Methods

The presentation of Byars et al.'s (2010) methods in section 7.2 is quite elliptical. Although it's not easy to provide a clear and brief summary that goes further, I have to try.[25] This will also allow me to clarify some of my claims in chapter 5 about how unmeasured factors get incorporated into Byars et al.'s analysis.

Byars et al. performed a multivariate linear regression using measurements of several other traits in addition to age at first birth (e.g., total cholesterol), and they controlled for possible changes in societal and environmental factors, as well as each woman's development over time, and other traits that were measured as part of the FHS. The authors also calculated regressions on interactions between traits, though these interactions did not explain much, and they used squared trait values to estimate stabilizing selection. Specifically, Byars et al. (2010, 1791) used a standard fitness regression model due to Lande (1979) and Lande and Arnold (1983), $\Delta z = G\beta = GP^{-1}s$, where Δz is a vector of average predicted change in phenotypes, s is measured changes in phenotypes, G is a matrix of additive "genetic" variance and covariances between parent and offspring traits, P is a matrix of measured variances and covariances between traits, and β is a vector of selection gradients.

More specifically, s measures a "within-generation" change in trait frequency by subtracting trait frequencies among mothers from the same trait frequencies weighted by each mother's LRS, without looking at trait frequencies among offspring. Δz reflects the change between trait frequencies among mothers and trait frequencies among offspring. P is assumed to be the sum of G (which reflects intrinsic additive heritable interactions) and other factors including pleiotropic and epistatic interactions and environmental influences. The "additive genetic variance" reflected in G could also involve nongenetic influences of mothers on their daughters, and s and P might also reflect influences from family members, friends, coworkers, broader cultural factors, jobs, economics, and so on. Note that although it's not obvious, one can view the calculation of this regression coefficient as an application of the Price equation in its simplified Price-Robertson equation form.

7.7 Appendix: Pigliucci and Kaplan on Lande-Arnold Models

In chapter 1 of their book, Pigliucci and Kaplan (2006) adopted a view of natural selection and drift that is primarily statisticalist, borrowing from works such as Matthen and Ariew (2002), though some passages suggest that their view may allow causal claims that most statisticalists would not allow. Chapter 2 of the same book is devoted to a critique of Lande-Arnold style regression methods, such as those illustrated in section 7.2.

All of Pigliucci and Kaplan's chapter 2 arguments against Lande-Arnold methods seem to have the same form.[26] Pigliucci and Kaplan argued that assumptions that empirical researchers make and the ways that researchers use quantitative genetic methods are not sufficient to justify the researchers' general causal claims about evolution in the populations they are studying. Such claims, like those about selection gradients in Byars et al.'s (2010) study, are general causal claims about what is driving evolution. (These are not, for example, claims about particular token organisms.) Pigliucci and Kaplan clearly assumed that there is a correct general causal story about what produces evolution in a population, and they argued that Lande-Arnold methods are inadequate in practice for learning about this story.

Initially, the problem with Lande-Arnold methods, according to Pigliucci and Kaplan, is that these methods are not able to reflect causal relationships in a sufficiently fine-grained manner. For example, Pigliucci and Kaplan wrote:

There is nothing in the more complex multivariate case that would permit an easy move from statistical associations to causal structures. But what we wish to learn by studying selection is the causal pathways between traits and differences in fitness—not just the statistical associations! (Pigliucci and Kaplan 2006, 53)

That is, for Pigliucci and Kaplan, one problem with Lande-Arnold methods is that they don't justify inferences about causal relationships between traits and fitnesses. The causal relationships can't be read off the statistics, they claim—not because that's impossible in principle to do so but because Lande-Arnold regression methods are not up to the task. At the end of the chapter, Pigliucci and Kaplan argued that the right way to investigate causal relationships between traits and fitnesses is by use of a related but more elaborate statistical method, path analysis (Wright 1934; Rice 2004). The thing to notice is that this view seems entirely consistent with my view about fitness and natural selection, which is that it concerns hidden causal factors that must be estimated from observations. As far as I can tell, Pigliucci and Kaplan's position on Lande-Arnold methods does not conflict with the conclusions I drew using the example of Byars et al.'s (2010) study.

However, though Pigliucci and Kaplan's arguments in chapter 2 start from an assumption that (as I would put it) causal type fitnesses exist and are worth investigating, their arguments could undermine my claim that examples like Byars et al.'s study are illustrations of widespread *successful* scientific practices. Let's consider that possibility. I'll focus only on the most challenging of Pigliucci and Kaplan's criticisms of Lande-Arnold methods. This is illustrated by their use of a quotation from Mitchell-Olds and Shaw (1987, 1159):

Mitchell-Olds and Shaw (1987) argued that using the resulting selection gradients in the *multivariate breeder's equation* in order to *predict the long-term results of selection* involves a number of assumptions; namely, that

(1) many genes contribute to genetic variances and covariances; (2) genetic variance-covariance matrices remain approximately constant through time, which implies weak selection and large population size; (3) new or unmeasured environments do not alter G, P, or z [matrices and a vector used in the methods]; (4) there is no genotype-environment correlation or genotype by environment interaction. In addition, inbreeding and nonadditive gene action may greatly complicate prediction of response to selection. (1159)

Given that these assumptions are *known generally not to be true*, it is hard to imagine a clearer exposition of the strict limitations of the evolutionary quantitative genetic research program. (Pigliucci and Kaplan 2006, 56–57; italics added)

What's puzzling to me, first, is that these constraints on uses of Lande-Arnold methods are also highlighted by careful advocates of use of the methods, such as Lynch and Walsh (2018). I take such provisos to be a standard part of advice concerning use of these methods. Second, prediction of long-term outcomes is not the only purpose of using Lande-Arnold methods, in which the multivariate breeder's equation is central. The methods can also be used to understand or predict short-term change (as Byars et al. do), where the assumptions that Pigliucci and Kaplan highlight are less problematic. Third, the mere fact that assumptions on which a modeling method is based are literally false is not a reason not to use the method (see §2.1). What matters is the degree to which the difference between what the model assumes about the world and the precise truth prevents drawing conclusions that may be probably approximately correct (Weisberg 2013; Potochnik 2017). This point is crucial to modeling in many areas of science, but it's not one that Pigliucci and Kaplan seem to address. It may well be that when Mitchell-Olds and Shaw wrote in 1987, there was a need to warn against misuses of Lande-Arnold methods, and I can easily imagine that there are still some misuses. I don't take this possibility to undermine my view that Byars et al. (2010) represents an illustration of common, successful practices in evolutionary biology.

Finally, it's clear from the last section of the chapter that Pigliucci and Kaplan think that Lande-Arnold methods give us mere statistical relationships, while path analysis has the ability to investigate causal relationships. That is, it's not just that path analysis allows investigation of more fine-grained causal relationships than Lande-Arnold methods. Pigliucci and Kaplan apparently don't think that coarse-grained relationships such as those represented as selection gradients are causal. However, I see no reason such coarse-grained relationships are not causal facts about a population-environment system as a whole—even if, given appropriate data, other methods are able to provide evidence for more fine-grained causal relationships. Perhaps Pigliucci and Kaplan's refusal to grant causal status to things like selection gradients is due to lack of the concept of a population-environment system, in which causal probabilities of later states conditional on earlier states allow for a variety of causal relationships. The fact that the idea of a population-environment system helps to make sense of a great deal of research based on Lande-Arnold methods, which evolutionary biology at least implicitly treats as investigating causes, is then a point in favor of the population-environment conception: Pigliucci and Kaplan argue that applications of quantitative genetics fail to identify causal relationships, so that a great deal of scientific practice must be misguided. I argue, by contrast, that this scientific practice is a reason to think that causal relationships assumed by it are real, and that the population-environment system conception allows us to understand this idea. As is often remarked, one person's modus tollens is another's modus ponens.

Chance in Population-Environment Systems

8.1 Introduction

Chapters 2 and 7 included arguments that evolution depends on causal probabilities of events that can occur in the operation of a population-environment system. Moreover, by arguing against the view that evolution depends on probabilities of outcomes for actual token organisms (chapters 3, 4, 7), I provided reasons to think that the causal probabilities relevant to evolution are realized by a population-environment system as a whole. This view helps us to understand, for example, what the causal type fitnesses discussed in chapter 4 and 7 are: they summarize causal probabilities of possible outcomes in developmental paths for organisms of a given type.

But what are these causal probabilities? What is their nature? How do they arise in population-environment systems? Is there an interpretation of probability (§0.2) that applies to population-environment systems? I've said very little that might suggest answers to these questions. Answers are indeed difficult to come by, and I won't try to give a definitive proposal here. However, after explaining why some of the more obvious answers are inadequate, I want to investigate one kind of possible answer to the questions in this chapter and the next: it is that what I now call "measure-map complex causal structure" (MM-CCS) probability, developed separately by several authors including me, might be realized by population-environment systems.

Note that any proposal about the nature of probabilities in evolutionary processes must be somewhat speculative, as previous proposals illustrate (e.g., Brandon 1978, 1990; Mills and Beatty 1979; Brandon and Carson 1996; Millstein 2003, 2016; Drouet and Merlin 2015; Strevens 2016). This is, I believe, due to the extreme lumpy complexity (§1.2) of population-environment systems. It's just not possible to provide a characterization of the relevant processes that is systematic enough to fully justify an interpretation of probability or some other solution. Nevertheless, one can try to argue that a particular proposal is plausible, and more plausible than alternatives. Because of the lumpy complexity of real population-environment systems, my discussion in this chapter will make use of idealized examples that seem far from biology. However, I'll indicate along the way how my examples illustrate properties that could be realized in population-environment systems.

I begin in section 8.2 with a discussion of ways that population-environment systems might not quite realize causal probability per se; this provides context for some later points. In section 8.3, I argue that population-environment probabilities are not single-case propensities, and that postulating that they are long-run propensities or Sober's "no-theory" probabilities does not provide sufficient illumination about probability in population-environment systems. Section 8.4 begins to develop the basis for a proposal that MM-CCS probability might be realized by population-environment systems. My discussion of MM-CCS probability continues in chapter 9, where I consider one of the main challenges facing such a proposal and describe a possible solution.

8.2 Must It Be Probability?

I remarked earlier that my arguments for the existence of causal probability in population-environment systems would have to be weakened. The reason is that any claim that there are objective probabilities in some real-world system asserts something very precise: that the system realizes probabilities that can be expressed with precise, real numbers—as all probabilities can—and usually that these numeric values will be somewhat stable over time. Further, a claim about causal probabilities means that each such number must quantify a sort of center of gravity for actual frequencies of outcomes, and this center of gravity must be manipulable by physical means—though always remaining numerically precise. My arguments for the existence of causal probability were based on successful uses of probabilistic modeling and statistical methods to make inferences about real population-environment systems. Yet these methods only provide approximate evidence for approximate conclusions, as I've noted. So the successful use of the models and statistics mentioned in these arguments is evidence only for something that is at least approximately like causal probability.

First, it may be that the chanciness of population-environment systems involves only what are known as "imprecise probabilities" (Augustin et al. 2014) or "indeterminate probabilities" (Levi 1980). *Objective* imprecise probabilities, or *imprecise chances*, are similar to objective probabilities but don't necessarily involve precise, real-numbered values.[1] Mathematically, imprecise probabilities satisfy axioms of various sorts that generalize standard probability axioms, and their values are sets of real numbers, such as intervals in [0,1]. *Causal imprecise probabilities* would then be imprecise chances in which relative frequencies tend to vary in ways that are like manipulations of imprecise probability values.[2] The facts about practices in evolutionary biology that I used earlier in the book to justify claims that there are causal probabilities in population-environment systems are consistent with the existence of mere causal imprecise probabilities. These must have imprecise values that are sufficiently constrained and close to precise probability values that modeling and statistical inferences that assume the existence of probabilities can be successful. Elsewhere (Abrams 2019), I argued that there are actually good reasons to think that population-environment systems often realize objective imprecise probabilities, but for this chapter, I merely want to allow the possibility that the chanciness of population-environment systems could involve causal imprecise probabilities. The imprecise chance of an organism with a particular trait having a certain number of offspring could, for example, be an interval, say [0.231,0.234], rather than a single precise number. This would imply, we can suppose, that among realizations of this trait, frequencies of numbers of offspring would usually be in the neighborhood of [0.231,0.234].

8.3 Putting Aside Some Alternatives

In this section, I explain why I think that some potentially obvious candidates for population-environment system chances won't do.

8.3.1 THE FAT-CHANCE ARGUMENT

In Abrams (2007a, 2012c), I gave versions of what I call the "fat-chance" argument (Abrams 2012c) concerning single-case *propensity* (§0.2.3). Here is the same argument formulated for causal single-case probability in general.

Consider a particular coin flick (§2.2.2). The coin will spin for a little time and then fall down, with the upward-facing side indicating the outcome. (If the coin manages to balance on its edge, one must spin again to determine the outcome.) This token flick is (a) a flick of a coin, (b) a flick of a coin with a slight weight bias toward heads, (c) a flick at a precise angular velocity, (d) a flick whose vertical spin axis leans slightly to the left, (e) a flick in which there will be some tiny puffs of wind pushing against the spin direction, and so on.

It seems as if each of these properties could determine different probabilities of outcomes. The chance of falling on heads might depend closely on the initial velocity of the coin, for example—but the other factors might matter too. Yet on a simple coin flick, one might expect that the chance of heads is close to 1/2.

It seems that chances—or at least causal probabilities—should capture some kind of causal relationship between a chance setup and an outcome. For a single-case chance, this must be between a particular token of a chance setup and an outcome. One might then want to say that the same token setup determines a chance of 1/2 for heads, in one sense, and a chance of some other value, in another. One might think, for example, as a realizer of all of the details of the token flick, the chance of heads should be 1, but as a realizer of the coin-flick property alone, the chance of heads on this token flick is 1/2.

However, there is only one instance of any particular token flick, and this flick realizes the entire set of conditions that are true of it. Claiming that exactly the same conditions—those realized in the token setup—determine causal probabilities with two different values for the same outcome is, roughly, to claim that the causal relation between the token setup and the (type) outcome has two different strengths. It is analogous to claiming that a token event both will and won't cause some other (type) event. Indeed, this last claim is close to saying that the chance of heads is both 0 and 1. Note that since we are talking about a particular token chance setup, it's irrelevant that we can specify the properties of the setup in different ways (as a realizer of a coin flick, as a realizer of a flick with a particular initial velocity, etc.) What is causing the outcome is the collection of all of the properties realized in that setup that might affect the outcome. So it seems incorrect to claim that the single-case chance of heads can have two different values. There is only one correct value. Which one? Clearly, it's the probability relative to the most detailed set of causally relevant factors. For most high-level physical setups, a given outcome will have either a chance near 0 or near 1, relative to all of the details of the token setup.[3] Thus for most higher-level trials and events (in which quantum mechanical chances work out to the extreme values 0 and 1), only probabilities near 0 and 1 turn out to be true single-case chances. This means that higher-level chances of the kind relevant to evolution must be long-run chances (see Abrams 2015). These are chances relative to a less than maximally detailed specification of a chance setup. What such chances affect are not outcomes on single trials, per se, but frequencies of outcomes in large numbers of trials. A long-run chance interpretation specifies what it is about such repeated realizations that make them, collectively, objective realizers of axioms of probability.

8.3.2 POPULATION-ENVIRONMENT CHANCE
ISN'T (JUST) PROPENSITY

Arguments in chapters 3 and 4 show that the probabilities underlying evolution are not, in general, single-case propensities (§0.2.3) of outcomes for actual token organisms. One might think, though, that we could understand the chances underlying evolution as single-case propensities for a population-environment system as a whole to exhibit various future states. Maybe a chance setup that consists of a population-environment system in an initial state at particular a time (§1.4.3) has a single-case propensity for all of the possible ways that a population-environment system might develop, consistent with that initial state and constraints imposed by external conditions specified by researchers (chapters 5, 6).

One problem with the proposal that population-environment system chances are single-case propensities is related to points I made in section 3.3.3: Nearly all of what matters in the development of a population-environment system consists of variations in higher-level processes involving organisms, rocks, dirt, internal organs of organisms, interactions between organisms, and so on. These are processes in which quantum mechanical effects usually work out to processes that are nearly deterministic. Even intracellular interactions seem to depend mainly on deterministic Brownian motion of molecules: quantum mechanical effects on these interactions are so small that they are difficult to calculate (Kuriyan, Konforti, and Wemmer 2013, chap. 6). So, while population-environment systems are enormously complex, for the most part their development is likely to be nearly like a deterministic process: for any given initial state of a population-environment system, exactly one developmental path is overwhelmingly probable. The fat-chance argument then implies that there are no other *higher-level chances* for outcomes of this token population-environment trial; the only relevant chance setup consists of a complex configuration of many low-level properties. This means that the causal probabilities that play a crucial role in successful empirical research in evolutionary biology can't be single-case propensities.

Alternatively, suppose that indeterminism does make a small range of developmental paths probable for a given token population-environment system. Given the complexity of population-environment systems, with various sorts of sensitivity to small variations in earlier states, it's possible that quantum mechanical effects *sometimes* make a difference to evolutionary outcomes in population-environment systems. For example, *E. coli* bacteria have receptors that tend to cause them to swim toward beneficial substances and away from noxious substances. These receptors are complexes of molecules that interact with other molecules moving chaotically in the fluid inside the bacterium. In general, the motions and interactions of these molecules seem to result simply from deterministic movements in a fluid, but perhaps in rare cases the motions are influenced by quantum mechanical chanciness (Abrams 2017). Suppose that because of this, there is some chance that an *E. coli* bacterium will "accidentally" move a bit less rapidly toward a beneficial substance, with resulting poorer health, a longer delay before cell division, and a subsequent decrease in its numbers of descendants at a later time. If quantum mechanical chanciness involved single-case propensity, then single-case propensity would thus have influenced the relative representation of a particular genetic pattern in the population at a later time. There might be similar possibilities of quantum mechanical influences on cell functioning in more complex organisms.[4]

I think such quantum mechanical influences on evolution are rare at best, and are rarely significant. However, even if one thought that quantum mechanical chances played a more significant role in what are largely deterministic processes, it's unlikely that chances in population-environment systems derive solely from quantum mechanical effects. If they did, that would mean that the complex processes in a population-environment system would have to be so structured that all variations affecting the numbers of offspring for a particular trait would be dependent on quantum mechanical differences. That seems to require a very particular sort of causal structure, in which no variation between circumstances for realizations of a trait depends solely or primarily on deterministic processes. Various arguments in section 3.3 suggest that this is not the case. Moreover, as we'll see, there are interpretations of probability that allow complexity in a deterministic process to give rise to the same sort of regular variation in relative frequencies of outcomes for a trait. I therefore think it's unlikely that population-environment chances are due primarily to quantum mechanical effects, and that they are not single-case propensities. In the rest of this chapter, I'll usually assume that population-environment chances are based on deterministic processes; it's not hard to add quantum mechanical chances to the story if need be (Abrams 2015).

What about long-run propensities (§0.2.3)? These are partial-strength dispositions to produce relative frequencies of certain sorts, but the dispositions don't concern the outcome of any one trial (Gillies 2000; Eagle 2004; Berkovitz 2015; Hájek 2019). There need be no assumption that these dispositions have their source in fundamental physics, and long-run propensities are usually viewed as compatible with determinism. So one might think of population-environment system chances as dispositions for population-environment systems to produce, say, organisms that realize a certain trait with numbers of offspring in certain frequencies, roughly speaking.

Perhaps it will turn out that causal type fitnesses depend on long-run propensities, but this bare claim by itself doesn't tell us anything about *how* population-environment systems give rise to long-run propensities. Without such an account, the source of population-environment chances would remain a mystery, and with such an account, it's not clear that calling chances "long-run propensities" would add much. A similar point applies to Sober's (2010) "no-theory" interpretation of probability, according to which probabilities exist when they play the right kind of role in well-confirmed theories.[5] We are better off looking for more specific proposals about the nature of population-environment chances, even if the proposals turn out to be vague in some respects.

8.3.3 COULD POPULATION-ENVIRONMENT CHANCES BE LONG-RUN?

My view is that because population-environment chance is higher-level chance, and the fat-chance argument implies that higher-level chance typically can't be single-case, population-environment chances must be long-run (though not necessarily long-run propensities).[6] This means, prima facie, that such chances can exist only when there are multiple trials. However, it might be thought that each population-environment system is unique, and that therefore no long-run interpretation of probability can give us the sort of chances needed to make sense of evolutionary biology. Moreover, each research project in evolutionary biology typically performs no more than a few studies of a single population-environment system, even if they continue to study the same population for years.[7] There are some reasons that such concerns can be put aside.[8]

First, note that my conception of causal probability (§2.2.1) allows frequencies of outcomes, which tend to parallel probabilities, to be frequencies of outcomes stemming from a single trial of a chance setup. Some of the analogies in my introduction of the idea of a population-environment system (§1.4.1) conveyed this point. Thus my view is that long-run chance doesn't require multiple, independent trials. The relative frequencies that tend to have values close to probabilities, on a long-run interpretation, could be frequencies resulting from the same, long-running, complex trial. From this point of view, even if there is only a single trial of each population-environment system, there are multiple outcomes from each population-environment system, and it is these outcomes that researchers study.

Second, the fact that the chance setup for a population-environment system is defined by only some of its properties, as specified in section 1.4.3, means that multiple instances of the same population-environment system could be realized by the same ongoing population and environment at different times. For similar reasons, different subpopulations or subenvironments could define different (token) population-environment systems if they shared the right properties. So it is at least theoretically possible for a long-run interpretation to apply to population-environment systems in virtue of the same chance setup being realized multiple times.

In addition, I argue in a paper in preparation (Abrams LongRun MS) that long-run chance interpretations of probability can be applicable even when the "repeated" trials are quite different. This argument depends on theorems like the general version of Kolmogorov's strong law of large numbers, which imposes only very weak requirements on the similarity of chance setups. The conclusion of the argument is that a long-run chance interpretation of probability can have implications for general patterns of outcomes across trials of fairly disparate chance setups, even if there is only one trial for each chance setup. My view is that such general patterns are precisely what evolutionary biology provides.

8.4 Measure-Map Complex Causal Structure Chance

In recent years, several authors and I have separately worked on what I now call "measure-map complex causal structure" (MM-CCS) interpretations of probability. These are inspired by Poincaré's (1912; [1907] 1968; [1908] 1999) method of arbitrary functions and von Kries's (1886) natural-range conception of probability. Most MM-CCS interpretations attempt to define objective probabilities (J. Rosenthal 2010, 2012, 2016; Strevens 2011; Abrams 2012c, 2012d; Beisbart 2016; Roberts 2016), but some MM-CCS interpretations are objective/subjective hybrids (Myrvold 2012, 2021; Gallow 2021). There's also a closely related body of relevant work that doesn't focus on defining an interpretation of probability as such.[9] Though I'm not committed to the view that population-environment chances must be MM-CCS probabilities of some kind, the idea is still well worth considering. My arguments concerning probability in population-environment systems are related to arguments that Strevens (2003, 2008, 2013, 2016) has given concerning evolutionary processes, but my starting point is different (see Abrams [2012d] for a similar strategy in a different context). Some slightly technical details are necessary for understanding how to apply MM-CCS notions to population-environment systems, but I'll try to introduce the ideas in an accessible way.

8.4.1 THE WHEEL OF FORTUNE*

In chapter 1, I suggested that population-environment systems are somewhat like complex pinball machines. In this section, I use a different example, a simpler kind of device in which, again, similar initial conditions lead to different outcomes.

A wheel of fortune is a simplified roulette wheel with no ball; the outcome is indicated by the color of the wedge near a fixed pointer when the wheel stops. If such a wheel is repeatedly given a fast spin, the frequency of an outcome in a set of such spins will usually be close to the outcome's proportion along the circumference of the wheel. For example, if the wedges are of equal size and alternate between red and black, the frequencies of red or black will usually be close to 1/2. Other configurations are possible, however. A wheel with black wedges that are twice as large as red wedges, for example, should generate frequencies of black of about 2/3. We can assume that this correspondence between frequency and size generally holds no matter who the croupier is. (Casinos depend for their livelihood on such regularities.) Despite the fact that a wheel of fortune is a deterministic system whose outcome is determined by the initial angular velocity with which it is spun, it's natural to think there is in some sense an objective probability of an outcome that's equal to proportion along the circumference. This seems reasonable even in unusual cases in which a run of good or bad "luck" produces a frequency that departs significantly from that probability.

The wheel of fortune is what I call a *causal map device*, a (nearly) deterministic device that, given an initial condition, produces an outcome (figure 8.1). A croupier gives the wheel a spin—imparts an angular velocity to it—and Newtonian physical processes take over. Note that we can divide the set of initial conditions, consisting of angular velocities that are possible for human croupiers, into those regions that lead to the red outcome and those that lead to the black outcome. Figure 8.2 provides a schematic representation of this point. Velocities are represented on the horizontal axis, increasing as we move from left to right. The horizontal line is divided into intervals of velocities leading to red or to black, and the wedge to which each interval maps is indicated schematically by curved lines rising vertically toward boundaries between wedges.

The wheel of fortune is also "bubbly": its input space can be divided into many "bubbles," small regions of similar initial conditions that include conditions leading to all outcomes—in this case, to red and black.[10] In the case

---

* This section contains brief excerpts from Abrams (2012d). Reprinted by permission.

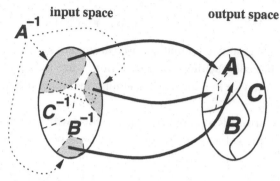

input space   **output space**

FIGURE 8.1. Causal map device: abstract structure. The oval on the right represents a space of possible output effects divided into three outcomes, *A*, *B*, and *C*. The oval on the left represents a space of initial conditions, divided up into sets of initial conditions, A^{-1}, B^{-1}, and C^{-1}, leading to respective outcomes *A*, *B*, and *C*. The diagonal dotted rectangle is a bubble (see text). Based on figures in Abrams (2012c, 2012d).

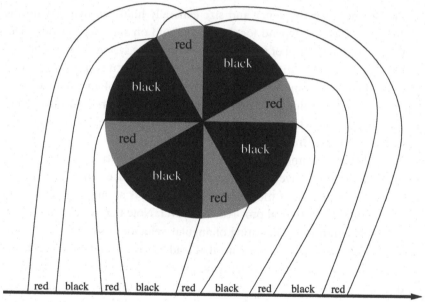

FIGURE 8.2. Wheel of fortune: velocity-to-outcome mapping. Originally published in Abrams (2012d).

of a wheel of fortune, consider any adjacent pair of intervals, one containing velocities leading only to red, and one containing velocities leading only to black. The union of those two intervals counts as a bubble, because it contains initial conditions, or inputs, leading to each outcome. Intuitively, a bubble must be small, in this case because the distance from its smallest to its largest velocity is a small fraction of the distance from one end of the input space to the other. Later we'll translate this assumption into a claim about probability measures.

We can plot the number of spins a given croupier imparts at any given velocity. Figure 8.3a illustrates this idea for a croupier who spins the wheel many, many times, but who tends to give the wheel faster spins more often than slower ones. The height of the curve at each point reflects the number of spins the croupier gives at the velocity represented by a point on the horizontal axis. This is a density curve (§0.2.1), since it represents the number of spins at each value of a continuous quantity (i.e., velocity).

It can be proven that if bubbles are small, and the proportion of red and black in each bubble is the same (more on this below), and, in addition, the slope of a croupier's density curve over inputs is not extremely steep in any region—that is, if the inputs are *macroperiodic*—then the frequency of red (black) will be close to the proportion of the input space occupied by points leading to red (black) (Engel 1992; Strevens 2003; Abrams 2012c; Myrvold 2021).[11] The relevant proofs provide at least a partial explanation of what it is about the wheel of fortune that gives rise to the phenomenon of stable frequencies that match proportions of the input velocity space. The general idea can be seen in figure 8.3a. Notice that since the height of the curve at a particular velocity represents the number of spins at that velocity, the area of the region under the curve above a particular interval represents the total number of spins in that region of velocities; larger areas represent more spins. Roughly, we multiply the width of a region by the average height of the curve over it to get the approximate number of spins at velocities in the region. Now, since in each pair of contiguous red and black regions (in each bubble) heights don't differ much, the ratio between the number of spins in the two corresponding velocity ranges will be close to the ratio between the widths of the two ranges. Since this is true for each such pair—each bubble—the ratio between the number of red and black outcomes will also be close to the ratio between the summed velocity intervals leading to red and the summed intervals leading to black. In this case, that ratio is close to the ratios between

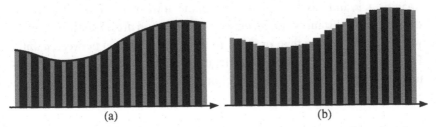

(a)   (b)

FIGURE 8.3. Hypothetical velocity distributions for a croupier. Based on a figure in Abrams (2012d).

sizes of red and black wedges.[12] Notice that it's not necessary that there be a smooth distribution over spins in order for this argument to work. A completely smooth distribution would require an uncountably infinite number of spins, anyway. All that's necessary is that each height—the total number of spins divided by the width of the velocity interval—be close to the same quantity for adjacent intervals, as in figure 8.3b. This is a histogram with bin widths corresponding to sizes of velocity intervals. Similar analyses can be given for other, more complex devices, such as coin- and die-tossing setups (J. B. Keller 1986; Engel 1992; Diaconis 1998; Strevens 2003; Roberts 2016), and it's arguable that this sort of pattern is a large part of what explains the stable frequencies found in many mechanical casino games.

Notice that so far I haven't mentioned any kind of objective probability in this section apart from relative frequency, which is not a kind of causal probability (§0.2.3, §2.2.1).

8.4.2 FROM METRICS TO MEASURES*

As I mentioned, several authors have explored the idea that we might use such facts about devices such as the wheel of fortune to define an interpretation of probability. If the result is an objective interpretation of probability, this would usually be a causal probability interpretation, since manipulating the device could both manipulate probabilities and relative frequencies of outcomes in a coordinated way. For example, manipulating the widths of wedges in a wheel of fortune would, intuitively, manipulate both probabilities and relative frequencies. To explain how we might be able to apply similar ideas to population-environment systems, some additional terminology will be useful. This will allow us to abstract from the concrete details of the wheel of fortune.

A *metric*, which generalizes the concept of distance, is any function on elements of a set that assigns a nonnegative number d to pairs of elements (velocities, in the case of the wheel of fortune) such that the triangle inequality holds for any three elements a, b, c: $d(a, c) \leq d(a, b) + d(b, c)$. More intuitively, imagine a, b, and c as three points on a plane: if we define $d(x, y)$ as distance, the sum of the distances $d(a, b)$, between a and b, and $d(b, c)$, between b and c, will always be greater than or equal to the distance $d(a, c)$ between a and c. For a function d to count as a metric, it need not correspond to spatial distance, though; it just has to behave enough like spatial distance to satisfy the triangle inequality.

---

* This section contains brief excerpts from Abrams (2012d). Reprinted by permission.

A *measure* is a generalization of the concepts of area or volume, rather than distance. A function is a measure if it satisfies two of the probability axioms stated in section 0.2.1: (1) and (2b). If a measure satisfies axiom (3) in section 0.2.1 as well, it's a probability measure. That is the only kind of measure that will matter to us, but I will often call probability functions "measures" in this chapter to emphasize their mathematical character. Importantly, although measure is a generalization of area or volume, it needn't depend on a metric. Of course, we often calculate areas and volumes by multiplying distances—as when calculating the area of a rectangle—and some measures, such as Lebesgue measure (see below), can be defined with the help of metrics. However, the axioms that define the concept of measure don't depend on a metric. This point will be important for understanding the possibility of MM-CCS chance in population-environment systems.

In my explanation of relative frequencies in the wheel of fortune, we had an input space, consisting of ranges of velocities, and an outcome space, consisting of the outcomes red and black. What we saw was that as long as croupiers' initial condition distributions don't bunch up spins in very small regions favoring one outcome or another—as long as the distribution is macroperiodic—the relative frequencies of red and black would correspond roughly to the widths of wedges, or more precisely, to the lengths of intervals in the space of initial condition velocities (figures 8.2, 8.3). Notice that putting things this way describes the input space in terms of a *natural metric*, a distance measure defined in terms of the nature dimensions of a physical quantity—in this case angular velocity.

However, we can generalize the ideas illustrated by the wheel of fortune by measuring bubbles in terms of a (probability) measure on the input space (and similarity) rather than a metric.[13] Let's choose velocities u_0 and u_1 that are respectively a little below and a little above velocities possible for a human croupier. Then $[u_0, u_1)$ is an interval containing all of the possible velocities.[14] We can capture the intuitions about the wheel of fortune by defining the measure of any velocity interval $[v_i, v_j)$ as the difference between the interval's largest and smallest velocities divided by the difference between the largest and smallest velocities that are humanly possible:

$$P([v_i, v_j)) = \frac{v_j - v_i}{u_1 - u_0}.$$

This defines a probability measure, since if we partition (§0.2.1) the input space $[u_0, u_1)$ into disjoint intervals $[u_0, v_1), [v_1, v_2), [v_2, v_3), \ldots, [v_k, u_1)$, the probability of the union $[u_0, u_1)$ of all of the intervals is $(u_1 - u_0)/(u_1 - u_0) = 1$, and the probability of each smaller interval is a fraction of that quantity determined by the relative length of the interval. Then the probability $P([v_i, v_j])$ of an interval containing only a single velocity will be zero, and we can define the probability $P(\varnothing)$ of the empty set to be zero as well. This is the core idea of what's called a "normalized Lebesgue measure" over the wheel's input space.[15] Normalized Lebesgue measure defines a probability *measure* in terms of a *metric*. Notice that so far this "probability" measure over initial velocities is *just* a mathematical probability function that happens to be defined in terms of some physical quantities. Relative to the sense of "probability" that we usually intend when talking about events in the world, this probability measure is no different from the "probability" measure I defined for portions of a page in section 0.2.2. Both are probabilities in a purely mathematical sense. I'll return to this point below.

We can now put the points made earlier about the wheel of fortune in terms of a probability measure over the input space. Any continuous region of the initial velocity space now has a mathematical probability. Given a particular croupier, the region also has a certain number of spins that fall within it. We can use this number of spins divided by the measure of the region to define a "probability-average" of the number of spins in the region. (In figure 8.3*b*, the heights of bars are such probability-averages.) So this is no longer *just* an average defined directly in terms of a metric; we are now stating it in terms of probability—that is, in terms of a measure. Then the main claim about a wheel of fortune is as follows:

As long as a croupier's spins are such that

the probability-average number of spins within each bubble don't change much from one adjacent interval to the next,

then

the relative frequencies of red outcomes will be similar to the summed mathematical probabilities of the velocity regions for velocities that result in red (likewise for black).

Though strictly speaking this is the same as the point made above in terms of the metric over velocities, now we've stated it using a (mathematical probability) measure rather than a metric. So the "average" number of inputs in a red or black region is no longer the number of inputs in the region divided by the *distance* between its maximum and minimum velocities; henceforth we treat it as the number of inputs in the region divided by the (normalized Lebesgue) *measure* of the region.

Consider figure 8.4, which can be viewed as a detailed version of a pair of adjacent red and black bands in figure 8.3*b*. (Red is represented by gray in figure 8.3*b*.) We *could* understand width in this figure as representing the length

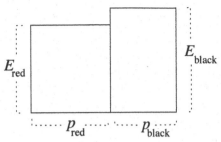

FIGURE 8.4. Input distribution within a bubble relative to an input probability measure. Areas represent total numbers of inputs in a region, width p_α represents measure of within-bubble region of inputs leading to outcome for $\alpha \in \{\text{red,black}\}$. Height E_α represents average within-bubble number of inputs leading to outcome α. Based on figures in Abrams (2012c, 2012d).

defined in terms of the natural metric on velocities, but since we've defined a Lebesgue probability measure in terms of this metric, we can treat widths as representing mathematical probabilities themselves. In the case of the wheel of fortune, we've simply made a terminological shift from averages defined in terms of a metric to averages defined in terms of a measure (which was defined in terms of the metric). The important point is that we now can do the same thing with input probability measures that are not defined in terms of metrics. It turns out, then, that the claims that justify the intuition that a wheel of fortune should usually generate frequencies corresponding to relative wedge sizes can now be made using a measure on the input space—even if there is no metric on the input space. (This will make it easier to justify applying MM-CCS ideas to population-environment systems, where it's implausible that relevant metrics exist.) In Abrams (2012c), I proved a theorem to this effect in terms of input probability measures for bubbly causal map devices (stated below in §9.5). Strevens's theorem 2.3 (2003, 136) and many of the theorems in Engel (1992) have similar conclusions, but those theorems depend on assuming that we have a metric over the input space.

One more point about the input measure: In the wheel of fortune, it's not necessary that all of the regions leading to black have the same input measure, nor that all of the regions leading to red have the same input measure. What is necessary is just that *within each bubble*, the proportion of the bubble that is black or red is the same as within each other bubble. It would be OK, for example, if some wedges of a given color were very narrow and others were wider, as long as within each bubble, the black wedge was twice as large as the adjacent red wedge (for example). To put it a different way, the probability of inputs leading to a given outcome, conditional on the bubble, must be the same for each bubble. In Strevens's (2003) terminology, the input measure must be *microconstant*. These points will apply to the general case when we move beyond wheels of fortune.

Where are we? So far, I've presented an intuitive picture that could explain why some devices such as wheels of fortune produce outcomes with frequencies that they do. I've started to generalize the basis of the picture, substituting a mathematical *input* measure for an *input* metric. What we really want in the end, though, are probabilities for *outcomes*. These must be more than merely mathematical probabilities: we need those probabilities to correspond, roughly, to the relative frequencies that are produced, in such a way that manipulating probabilities tends to manipulate frequencies. Moreover, we need to see how it might be possible to extend this picture to population-environment systems. The next two sections will move us toward achieving these goals.

8.4.3 BUBBLINESS AS A SOURCE OF CHANCE?

Suppose we go one step further now and define an *outcome* probability of black as equal to the input "probability" of the velocities that lead to black—and similarly for red. Call this the *mapped probability* for outcomes. In terms of the causal map concept, we are defining the probabilities of outcomes in terms of the probabilities of sets of initial conditions that lead to them. We are still assuming that the input probabilities are microconstant. Then since within each bubble the relative measure of inputs leading to each outcome is the same and the relative measures of results of those inputs is determined by their input measures, the overall measures of the outcomes must equal to the same proportions. For example, if within each bubble, black wedges are twice as large as red wedges, then the outcome probability of black is 2/3. But remember, the input probabilities are, so far, merely mathematical probabilities. This means that the outcome probabilities are merely mathematical too. In this respect, they are again like the page region probabilities defined in section 0.2.2.

However, we are now part of the way to a definition of an interpretation of probability that could be applicable to population-environment systems. (We might not get there!) In the case of the wheel of fortune, we've defined a probability measure over outcomes—red and black—in terms of physical quantities (velocities) in such a way that, as long as a croupier's spin distribution remains macroperiodic, we will have at least part of what is provided by causal probability—for we can manipulate relative frequencies of red and black by manipulating relative wedge sizes, and at the same time, we are also manipulating the probability measures (via Lebesgue measure) in the velocity space. For example, if we change the ratio between black wedge sizes and red wedge sizes, we cause corresponding changes in the ratios between input velocity interval widths, and between input probabilities. We also change which spins produce which outcomes. (But this is still just mathematical probability.)

As long as these changes in wedge sizes preserve macroperiodicity, the argument that relative frequencies will be close to outcome probabilities still applies. On the other hand, these claims are dependent on the existence of a particular distribution of spins that happens to be macroperiodic relative to a particular velocity-defined Lebesgue measure. This assumption feels reasonable given what we know about things like roulette wheels. For less familiar, complex systems such as population-environment systems, analogous intuitions might be unwarranted. Nevertheless, this example illustrates some of the core ideas that advocates of MM-CCS interpretations have tried to build upon.

The crucial aspects of the story so far: The wheel of fortune is a causal map device in which the input space contains "bubbles"—subsets of small *measure* containing similar initial conditions, including conditions leading to each of the possible outcomes (e.g., red and black). Further, we've assumed—thus far— that sets of initial conditions that occur in the world—for example, from a single croupier—have frequencies that are macroperiodic: their average numbers don't change very much from a part of a bubble containing inputs leading to one outcome, to another part containing inputs that lead to another outcome. Finally, this macroperiodicity is relative to the measures assigned to the subsets of the bubble that lead to the different outcomes. That is, the sense in which the distribution of inputs doesn't change much is that there are similar average numbers of inputs relative to the measures of different parts of each bubble. After the initial definition of an input probability measure, a metric plays no role in this picture. (For further details see Abrams 2012c.)

Before discussing a deep problem concerning the input probability measure, I want to clarify the ways in which a population-environment system might be analogous to a wheel of fortune.

8.4.4 BUBBLY POPULATION-ENVIRONMENT SYSTEMS

My strategy for understanding the idea that evolutionarily relevant chances might derive from an MM-CCS picture will treat a population-environment system as analogous to a wheel of fortune. This differs from Strevens's usual strategy (2003, 2008, 2013, 2016; see also Strevens 2005), which treats part of the life of an organism in its environmental context as analogous to a wheel of fortune. After motivating my approach, I'll indicate why I see Strevens's conception as too limited.

The house sparrows example in chapter 1 and my discussion in chapters 3 and 4 were intended to convey the enormous complexity and subtlety of influences on the success of organisms that realize an inheritable type. These complexities include the following:

- Effects of both longer-term and transient variation within an environment
- Effects of interactions between traits within an organism—possibly with extreme phenotypic variation in different organisms due to sign epistasis (§4.5.3)
- Effects of spatial and other physical relationships between organisms that compete for food and shelter, provide sources of useful information or mutual aide to each other, or are related as sexual partners, predators, prey, parasites, hosts, or simply intervening bodies that impede movement

Because of the resulting subtle, lumpy complexity, I suggest that, in practice, for nearly all real-world population-environment systems, similar states at a given time often lead to evolutionarily different outcomes (see §3.3.3). In other words, developmental paths (§4.2.3) that include similar configurations of states at one time often result in particular inheritable types or circumstances leading to very different outcomes. (The Pinball Evolution machine analogy in chapter 1 was meant to suggest this idea.)

If that's correct, then it's plausible that there are reasonable ways of characterizing the possible initial states of the population-environment system that would make the system bubbly relative to those outcomes that an actual population-environment system would realize (§1.4.3). The idea is that it's possible to partition (§0.2.1) the initial condition space of a population-environment system into subsets with initial conditions leading to each member of a set of specified evolutionarily relevant outcomes such that

1. conditions in those initial condition subsets are more similar to each other than to conditions in other members of the partition, and
2. these subsets would be, intuitively, small compared to the entire space of possible initial conditions.

Then the partition members would be bubbles—at least if we can clarify the sense in which the subsets are small.

For example, if a population-environment system for house sparrows began at a date in the spring of a particular year, the preceding claim would be that there is a way to partition the circumstances in which house sparrows might be found (given the way in which the assumptions about the population-environment system constrain its possible initial states, §1.4.3) such that house sparrows beginning in similar circumstances end up with very different numbers of viable descendants, say, three years later. It's not necessary that scientists be able to construct such a partition; I doubt that anyone could. What matters is whether there is such a partition—since the question we are considering is whether the probabilities realized in population-environment systems are in fact MM-CCS probabilities.

In order for an MM-CCS interpretation of probability to apply to a population-environment system, additional requirements would need to be met (see chapter 9), including a rationale for the above sense of "small" in terms of an input measure. If the requirements plausibly were met, an MM-CCS interpretation of probability might provide an account of chances in evolutionary processes that could illuminate the basis of research in evolutionary biology. However, the existence of a metric on which a Lebesgue measure can be defined is not required by this framework (see §9.2.2). That's good, since the idea that there is a metric—the right kind of metric—on regions in the enormously complex space of possible initial condition circumstances for a population-environment system would seem very difficult to justify. Population-environment systems are not like wheels of fortune in this respect. (We'll see that there's a problem about justifying an input measure for population-environment systems too, but that's a different problem.)

Strevens's (2003, 2008, 2013, 2016; see also Strevens 2005) applications of MM-CCS ideas to evolutionary processes often treat each organism, or behavior, decision, interaction, or time segment in the life an organism, as a system to which MM-CCS properties such as macroperiodicity apply. That is, for Strevens, what is analogous to a wheel of fortune is an aspect of an organism in its environmental context. This is a natural strategy given the traditional focus of philosophers on causal token fitness, but it could be problematic for some of the same reasons I gave in chapters 3 and 4. That is, if one were to treat a *token* organism at a time and place as analogous to a wheel of fortune, that could make the MM-CCS chances of producing particular numbers of offspring too dependent on differences in individual circumstances, regardless of an organism's traits (§3.3.3, §3.3.4). This is the problem that the propensity interpretation of fitness had, translated to a MM-CCS context.

However, in a recent paper, Strevens (2016) stated explicitly that his MM-CCS account is to apply to organism types—for example, as I would put it, to realizations of alternative, competing traits in a particular population and environment. The problem with this idea is that evolution doesn't depend only on probabilities of outcomes for organism-type realizations in a fixed context. As I argued in section 4.4, the probabilities of success for possible realizations of an organism type also depend on probabilities of the type being realized in various environmental contexts and genetic backgrounds. Even if we put aside the problem of hitchhiking (§3.3.5, §4.4), fitnesses of types can't, in general, depend only on MM-CCS chances that are specific to organism types without taking into account chances of combinations of environmental and genetic circumstances for realized organism types. So, for Strevens's approach, we would also need a separate MM-CCS account of chances for circumstance variations. However, it's difficult to see how there can be an MM-CCS account of environmental circumstance variations that's not also an MM-CCS account of chance for the entire population-environment system. After all, organism trait realizations are among the things that affect downstream environmental circumstances for other organisms. Thus MM-CCS chances for traits will have to depend on the MM-CCS chances for the entire population-environment system. We should conclude that it must be population-environment systems, not organisms, that are analogous to wheels of fortune.[16]

Finally, although I'll continue to talk about population-environment systems themselves as realizers of MM-CCS properties, for cases in which a population-environment system is defined partly in terms of probabilities of external influences, as in sections 5.4 and 7.2 or section 6.4, in order to understand the source of all of the causal probabilities influencing the population-environment in terms of MM-CCS properties, we may need to see it as embedded in a larger system that realizes such properties, thereby generating probabilities for external influences on the population-environment system.

8.5 Conclusion

I argued in earlier chapters that population-environment systems realize causal probabilities. In the first section of this chapter, I suggested that this claim should be weakened to allow for the possibility that population-environment systems realize causal imprecise probability. I then argued that population-environment probabilities are not single-case propensities, and that claims that these probabilities are long-run propensities or no-theory probabilities would tell us very little. I began to outline the main ideas of measure-map complex causal structure (MM-CCS) interpretations of probability and how to apply them to population-environment systems. I emphasize that at this point, the account on the table so far seems to define MM-CCS chance in terms of a purely mathematical, potentially arbitrary input measure. The next chapter concerns the possibility of addressing this problem.

The Input Measure Problem for MM-CCS Chance

9.1 Introduction

The previous chapter introduced basic ideas of measure-map complex causal structure (MM-CCS) chance and outlined my strategy for applying these ideas to population-environment systems. At this point, the account is very incomplete, though. Probabilities of outcomes have so far only been defined by a potentially arbitrary, purely mathematical input measure. This chapter is devoted to clarifying the challenge this idea raises (§9.2), and to investigating a potential response to the challenges (§9.3).

9.2 The Input Measure Problem

The primary challenge for understanding causal probability in population-environment systems as a kind of MM-CCS chance is this: we need to make it plausible that for a given set of evolutionarily relevant outcomes, there is some way of justifying a measure over allowed initial states (chapters 1, 5, 6) of a population-environment system such that

1. there are bubbles with small measure;
2. different parts of these bubbles leading to different outcomes have the same (or close to the same) conditional probability relative to each bubble (the measure is "microconstant"), where this input measure in some sense captures facts about patterns of initial states of a population-environment system; and
3. these conditions hold in ways that fit with generalizations, predictions, and empirical research in evolutionary biology.

The MM-CCS *input measure problem* for population-environment systems is the problem providing a justification for a choice of an input measure satisfying at least requirements (1) and (2). An MM-CCS input measure for a population-environment system should also contribute to understanding requirement (3). As suggested by various remarks above, no detailed specification of an input measure satisfying these requirements is likely to be possible for something as lumpily complex (§1.2) as a population-environment system, but it would be good to be able to sketch a story that made it plausible that such an input measure existed. There is an analogous problem for any MM-CCS interpretation of probability, so I'll again make use of the simple example of the wheel of fortune to help to clarify ideas. Note that any account of chance for evolutionary processes has to satisfy a requirement analogous to (3), but no previous proposals have done so (see citations in sections 8.1 and 8.3).

9.2.1 ARBITRARY INPUT PROBABILITY MEASURES

In section 8.4.2, I defined an input probability measure for a wheel of fortune as a normalized Lebesgue measure with respect to initial velocities, and then I assumed that relative to that measure, croupier spin distributions would be smooth—macroperiodic—in the sense illustrated by figure 8.3a, or by figures 8.3b and 8.4. It seems as if this scheme defines probabilities that tend to match frequencies of outcomes, and that altering wedge sizes alters outcome probabilities in ways that would parallel changes in outcome frequencies. These are thus causal probabilities. If the assumptions about input measure and outcome frequencies for a wheel of fortune are correct—and nearly everyone who writes about MM-CCS interpretations thinks they are—we should still be very puzzled. Why would one think that *that* measure—of all of the uncountably infinite options—would give us causal probabilities when human croupiers spin wheels of fortune? Some clarification of the question may be helpful.

Consider the lower of the two small curves in figure 9.1a. Intervals of velocities leading to black or red are marked with vertical lines. If we assume that we can represent croupiers as having continuous initial velocity distributions (as in figure 8.3a), then this lower curve could represent part of one croupier's velocity distribution when the wheel was spun by hand in the normal way. Suppose, though, that sometimes the same croupier operates the wheel by moving an arm on a mechanical device that causes the wheel to spin. The croupier transmits the same force to this arm as would have been imparted spinning wheel by hand, but the levers and gears in the device cause more spins in certain velocity regions and fewer in others. This produces the partial spin distribution represented by the upper curve in figure 9.1a. (Since there are more spins in this region, there will be fewer in other regions not

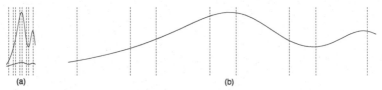

(a) (b)

FIGURE 9.1. *a*, *Lower curve*: croupier's natural velocity distribution over a portion of a range of allowed velocities. *Upper curve*: velocity distribution with the aid of a mechanical arm (see text). Vertical lines delineate velocities on the x-axis that lead to red and black. *b*, The same mechanical-arm distribution as in 9.1a but with greater measure assigned to each region.

represented in the figure.) We can see that with the mechanical arm, the croupier's distribution is no longer macroperiodic: there are large jumps between adjacent intervals, and the ratios between numbers of red and black spins within some bubbles are no longer close to ratios between velocity intervals. We can no longer expect that croupier's spins will produce outcomes with frequencies close to probabilities as defined by our normalized Lebesgue measure on velocities. Thus, the fact that probabilities of outcomes can sometimes be understood in terms of Lebesgue measure, even for a simple device such as a wheel of fortune, should be viewed as a special circumstance due to facts about croupiers and the design of wheels of fortune.

Suppose now that we change the input measure from the one in figure 9.1a to the one in figure 9.1b, where the velocity regions in the displayed region are assigned a larger input measure. I represent this in the figure by making the velocity regions wider in figure 9.1b, increasing the distances between vertical lines, while preserving the ratios between the distances. The curve displayed there is in fact the upper curve from figure 9.1a, but "stretched out." This makes what was the upper curve—due to use of the mechanical arm—appear more macroperiodic, and more like the original lower curve. However, we have now increased the probability measure of this region of the velocity space, and we must therefore assign smaller measure to other parts. In those regions, we may find that the mechanical-arm curve has become steep in places. Furthermore, although the curve for spins is more macroperiodic in figure 9.1b, the red and black regions no longer have small measure. As a result, despite the macroperiodicity of the curve per se, the differences in numbers of spins between (adjacent) red and black regions within each bubble is the same as it was for the upper curve in figure 9.1a.

To help understand what's going on here, consider figures 9.2a and 9.2b. The first of these is a version of figure 8.4. Recall that figure 8.4 showed a croupier's distribution of actual initial conditions leading to red and black

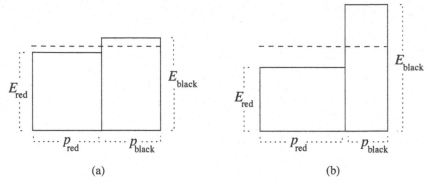

FIGURE 9.2. Input distribution within a bubble relative to two input probability measures. Areas represent total numbers of inputs in a region, width p_α represents measure of within-bubble region of inputs leading to outcome $\alpha \in \{red,black\}$. Height E_α represents average within-bubble number of inputs leading to outcome α. (See note 8 for the meaning of horizontal dashed lines.) Based on figures in Abrams (2012c, 2012d).

within a single bubble. Figure 9.2*b* represents exactly the same spins but displayed using a different input measure. The areas of the left rectangles in both figures—representing numbers of red spins—are in fact the same, as are the areas of the right rectangles, which represent the black spins in this bubble. The heights of the rectangles differ because of the difference between the values of the input measures of the two regions. This illustrates the idea that whether a distribution of actual initial conditions is macroperiodic depends on the measures assigned to regions in the input space. An input measure can't be chosen arbitrarily. There must be some reason that it's the right measure. Otherwise, a bubbly causal map device such as a wheel of fortune won't tend to produce frequencies that correspond to the mapped probabilities (§8.4.3) for outcomes.

9.2.2 THE PROBLEM

A primary problem for any MM-CCS interpretation is how to choose and justify the assignment of a measure to the input space for particular kinds of causal map devices. We want a way of choosing a mathematical probability measure on the initial condition space such that

1. within each bubble, subsets leading to each of the possible outcomes get the same measure (microconstancy), and
2. most of the time, perhaps for some good reason, average frequencies of initial conditions within a bubble that lead to different outcomes will be similar, relative to the measures of these subsets of the bubble (macroperiodicity).

At this point, it may sound as if no work would be done by bubbliness, microconstancy, or macroperiodicity. After all, it seems that we're just trying to find an input measure such that mapped probabilities match relative frequencies. Why not just use a frequency interpretation of probability directly (§0.2.3)? Or we could use some other interpretation of probability to define the input measure—but then aren't we simply trying to find a way to define probability in terms of another kind of probability, as some of de Canson's (2022) critical remarks suggest? The real work would then be in finding an interpretation of probability for the input measure, not in any of the rigmarole about bubbliness and so on.

However, bubbliness makes the determination of an input measure less of a problem than it could be. Bubbliness means that any probability distribution over initial conditions that preserves microconstancy and macroperiodicity is equally good. And for any given measure that does that, many others do as well. For example, for a wheel of fortune, any input measure that shrinks or expands bubbles in such a way that distribution curves for human croupiers continue to have gentle slopes will still produce outcome frequencies that are close to input probabilities. So, though bubbliness and other machinery described in chapter 8 are not in themselves enough to give us an interpretation of probability, an interpretation of probability based on MM-CCS ideas could be easier to justify than one without them.

9.2.3 PROPOSED SOLUTIONS

Existing MM-CCS interpretations of probability differ primarily in their responses to the input measure problem. I'll mention some of the options and briefly describe my reservations about them.

Jacob Rosenthal (2010, 2012, 2016) advocated basing input measures for some physical devices and systems on Lebesgue measure derived from natural metrics (§8.4.2). Strevens is more cautious but can be reasonably interpreted in some contexts as advocating similar ideas (Strevens 2003, 2011, 2013). Lebesgue measure based on natural metrics does seem as if it could provide the basis for an interpretation of probability for common physical games of chance (e.g., J. B. Keller 1986; Engel 1992; Strevens 2003).[1] However, as I argued in section 9.2.1, it's not clear why Lebesgue measure over natural metrics works in these cases, and it's even less clear why it should work in general. Strevens (2003, 2013) has given what he calls the "perturbation argument" that noisy microscopic variations in natural physical quantities should often make natural-metric Lebesgue measures close to the distribution of these quantities. This argument and the issues it raises are complicated, but the argument depends on assuming that only certain specific classes of noise tend to affect physical values in small ways. I don't think that Strevens's arguments for this assumption have been sufficiently convincing, but that is a topic that deserves a more detailed discussion than I can give here.

In order to directly apply the Rosenthal/Strevens Lebesgue measure idea to provide an input measure for population-environment systems, we'd first need a metric over the input space for a population-environment system. It's not at all clear that there are relevant metrics available for an application of MM-CCS to population-environment systems. For example, Lewontin (2000) describes a number of nonlinear developmental effects. What are the "natural metrics" for variation in physiological states of a developing fetus at the start of a population-environment system? Or consider the fact that nonlinear effect curves of protein concentrations are thought to be a reason for the dominance and recessivity relationships for combinations of alleles at a genetic locus (Gillespie 2004). What is the appropriate metric for such protein concentrations? These questions concern only simple variables, but the range of variations in initial conditions for a population-environment system is enormously complex. What would the right metric be, for example, for variations in the structural integrity of birds' nests due to interactions between nestlings' behavioral dispositions and available nest material?

What I'll call the *percolation strategy* was introduced by Strevens (2003) and advocated by Jacob Rosenthal (2012, 2016). It might make it possible to define MM-CCS chances for complex systems such as population-environment systems *if* there were a rationale for Lebesgue measure over natural metrics. The idea is to define MM-CCS chances over low-level physical variations with natural metrics, where these low-level chances are then chances of inputs to a complex system with higher-level effects. The higher-level chances thus derive from the low-level chances. For example, suppose there were Lebesgue-measure-relative MM-CCS chances of positions and velocities of molecules, and that variation in these molecules affected cellular processes or movement of microorganisms, while the latter processes in turn affected the functioning of organisms, and so on. Then, the proposal goes, the chances of the higher-level processes would depend on the original low-level MM-CCS chances. The low-level MM-CCS chances "percolate up" to the higher-level chances. This proposal has some of the standard input measure problems, but in addition, it depends on assuming that there is some chain (or directed graph) of mechanisms or devices that map lower-level MM-CCS chances into higher-level outcomes. For population-environment systems, it would be extremely challenging to justify such an assumption. (On the other hand, one can view this problem as inherent in trying to understand chance in a lumpily complex system.)

Roberts (2016) has suggested that natural laws over initial conditions could provide the basis for input probabilities. He focused on examples from physics and did not discuss lumpily complex systems. I don't think there are very good reasons to think that such laws exist, and I would be particularly skeptical that there are natural laws over initial conditions for population-environment systems (see above). On the other hand, if Roberts were right, the MM-CCS percolation strategy might allow Roberts's idea to apply to population-environment systems. The questions raised in the preceding paragraph arise here too.

Myrvold (2012, 2021) and Gallow (2021) argue that we can define input measures in terms of hypothetical Bayesian epistemic probability distributions that are "reasonable" in special senses. Myrvold and Gallow argue that this can be done in such a way that the distributions reflect objective facts and justify conclusions about relative frequencies. I am not convinced, however, that these distributions are sufficiently tied to objective characteristics to be able to provide the basis for causal probability, especially for something as complex as a population-environment system. It's also unclear how to apply Myrvold's or Gallow's ideas about reasonable epistemic probability to initial conditions of something that has the kind of complexity that a population-environment system has. The questions are similar to those I raised above about how to define metrics over initial conditions for a population-environment system: what would "reasonable" epistemic probabilities for initial conditions of such a complex system be?

Strevens (2011) and I (Abrams 2012c, 2012d) proposed MM-CCS interpretations of probability that define input measures as a function of certain actual frequencies of initial conditions. In my "mechanistic probability" interpretation, these were frequencies of initial conditions in a wide range of actual, similar systems. For a wheel of fortune, the systems would include other wheels of fortune and mechanical one-armed bandits. For a population-environment system, the similar systems could include the same population and environment at other starting times, or certain other population-environment systems, perhaps using ideas discussed in section 8.3.3. (In Abrams [2012d], I advocated similar ideas for probability in social sciences.) This proposal thus defines probabilities of outcomes for one actual causal map device in terms of frequencies of initial conditions to many actual causal map devices. Such an interpretation of probability is partly based on actual frequencies, but it doesn't share all of the flaws of a simple actual frequency account of probability. For example, I argued that the probabilities for outcomes of each such causal map device will usually be close to relative frequencies among its outcomes, though these probabilities and relative frequencies needn't be equal. Arguments in Abrams (2012c) imply that my interpretation was a causal probability interpretation: manipulating probabilities by manipulating the causal structure of the device—for example, changing wedge sizes in a wheel of fortune—tends to manipulate outcome frequencies in a parallel fashion.

However, Jacob Rosenthal's (2016) arguments convinced me that my interpretation, and Strevens's similar one, both have a deep flaw due to their basis in actual frequencies. These interpretations imply that a possible world in which frequencies are far from probabilities for most appropriately similar causal map devices is a world that is impossible *as a matter of logic*. Since input measures, and hence mapped probabilities, are defined in terms of actual initial condition frequencies across a large number of similar devices, mapped probabilities could never be far from frequencies for more than a few such devices. Even if such a world seems intuitively unlikely, it should not be built into a concept of probability that frequencies must be near to probabilities as a matter of necessity.

Strevens (2011) also proposed another, related, interpretation of probability that doesn't clearly suffer from the same problem, but it depends on assumptions about nearby possible worlds that seem unwarranted, and which I criticized in Abrams (2012c). See J. Rosenthal (2012), Beisbart (2016), Roberts (2016), and de Canson (2022) for other criticisms of MM-CCS interpretations of probability.

9.3 Bootstrapping MM-CCS with Stepwise Compression

There is an idea that I believe might possibly capture the true basis of an input measure for MM-CCS chances in population-environment systems. It is quite speculative, and difficult to motivate in detail. It shares some of the challenges of the percolation strategy, but it depends on weaker assumptions in some respects.

9.3.1 INITIAL ASSUMPTIONS

Let's start by supposing that there are various sorts of fluctuating patterns in, say, molecular motions, or wind gusts, or ways that pebbles bounce about on a hillside, and that these events provide initial conditions for some sort of bubbly-ish device (or devices) in nature. The causes of the physical variations are not particularly important. The device is only "bubbly-ish" because although there may be sets of similar initial conditions that lead to each member of a set of outcomes, these sets can't be considered small unless there is an input probability measure. I'll come back to that point. The patterns in the events-that-then-become-initial-conditions need not be anything like what we would expect from chancy processes. There needn't be any kind of causal probability distribution over them. The patterns might fluctuate wildly from one period to the next, and then settle near some value after a while, before shifting to an entirely different average value, and then varying in some new extreme manner. I won't even assume that the behavior of the initial conditions is systematic enough that it can be considered the result of *imprecise* chance (§8.2): the behavior might be "erratic"—that is, not chancy at all (Hájek and Smithson 2012; Abrams 2019).

What I'm going to argue is that if one assigns a fairly arbitrarily chosen mathematical probability measure over the input space of the bubbly-ish device(s) so that there are indeed small bubbles relative to this measure, then even if the wildly noisy input behavior that I described a moment ago is not macroperiodic relative to this measure, the device's actual outcomes can become more constrained and systematic, and more like outcomes of chance trials than the patterns in the inputs. What I'll suggest is that if this sort of process iterates, in that the outputs of each member of the first set of bubbly devices provides inputs to a second set of natural bubbly devices, outcomes from this second set of devices might as a result behave in ways that are closer to outcomes of chance trials. If *these* outcomes then become inputs to a third bubbly device, its outcomes may be even more like outcomes of chance trials—and so on. The end result of this kind of iterative process could be behavior in nature that truly is chancy, or that at least can usefully be modeled as chancy. (I'll ultimately leave it open to whether an interpretation of probability can or should be defined in terms of these patterns.) The core idea here is that an MM-CCS style device such as a wheel of fortune compresses, in a sense, variation in its initial conditions. To clarify this kind of *stepwise compression* process, let's look at an extremely idealized illustration.[2]

9.3.2 A SINGLE STEP

Consider a causal map device with many basic outcomes—that is, outcomes that can't be further subdivided.[3] Instances of these outcomes might then determine initial conditions for one or more further devices. For example, if the further device was a wheel of fortune, an instance of each of the outcomes from the first device could cause an initial velocity in a small region in

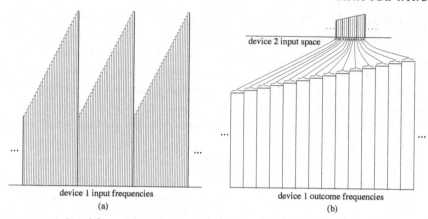

FIGURE 9.3. Partial frequency distributions. Dimensions are suggestive and not to scale. *a*, Idealized "bad" initial condition frequency distribution in three bubbles with thirty-two outcomes. *b*, Partial outcome frequency distribution, with causal links to region in a second device's input space.

the wheel's input space. Figure 9.3*a* illustrates this idea by representing three bubbles (out of many) for a device with thirty-two basic outcomes, each assigned the same input measure and represented within bubbles in the same order from left to right. Assume that the idealized initial condition distribution displayed for the three bubbles is repeated across many bubbles.

The initial condition distribution displayed in figure 9.3*a* is certainly not that of an MM-CCS system that generates frequencies near probabilities (outcome probabilities mapped by the input measure suggested by the figure). First, numbers of initial conditions within each bubble—represented by areas of narrow vertical rectangles—differ quite a bit between outcomes: We can see that because within each bubble the height over the rightmost outcome is more than twice as high as the height of the leftmost outcome. Thus the input distribution is not macroperiodic with respect to the displayed input probability measure, and relative frequencies for the thirty-two outcomes will be far from probabilities mapped to outcomes. (In the figure, the mapped probabilities of outcomes are identical—this is what's represented by the equal widths of the narrow rectangles.) Second, it's worth remarking that in most discussions of MM-CCS systems, authors illustrate the distribution over initial conditions with a curve that rises and falls, as in figures 8.3 and 9.1. When an actual distribution of initial conditions has that sort of shape, an excess of inputs leading to one outcome (e.g., red) in some bubbles tends to counteract a deficit of inputs for the outcome in other bubbles. That idea isn't essential to MM-CCS explanations of patterns in frequencies, but note that the distribution represented in figure 9.3a is a worst case, since the same outcomes get an excess or deficit of inputs in every bubble.

The crucial point about the nonmacroperiodic distribution illustrated in figure 9.3a is that though relative frequencies of outcomes won't be very close to the probabilities mapped from the displayed input probabilities, the bubbly causal structure of the device will cause outcome frequencies to be *more* like mapped outcome probabilities, as compared to (conditional) input frequencies and (conditional) input probabilities within each bubble. A partial outcome plot, labeled "device 1 outcome frequencies" in the lower region of figure 9.3b, can be constructed by reordering narrow rectangles from figure 9.3a so that rectangles with the same height are placed next to each other and are then merged into a single broad rectangle. This simple merging works because in this model each outcome has a characteristic within-bubble average frequency—that is, rectangle height; see figure 9.2. (In the general case, the height of each rectangle in figure 9.3b would be a weighted average of heights from the source rectangles, with relative widths as weights.) As a result, the angle of the staircase shape in figure 9.3b is less than the angles in figure 9.3a. This represents the fact that relative frequencies of outcomes displayed in figure 9.3b are closer to probabilities than within-bubble conditional relative frequencies are to corresponding within-bubble conditional probabilities in figure 9.3a. Bubbly causal map devices always "compress" relative frequencies like this, in that outcome frequencies are closer to mapped outcome probabilities than input frequencies are to within-bubble input probabilities. This point follows from a *compression theorem* proved in Abrams (2012c) and summarized in the appendix to this chapter, section 9.5.

9.3.3 ITERATING THE STEPS

Suppose now that instances of the outcomes from the device represented in figure 9.3 (device 1) are often inputs to a second bubbly causal map device, device 2, and that different device 1 outcomes produce device 2 inputs in different bubbles. The (fairly arbitrarily chosen) input measure for device 1 induces the mapped probabilities for its outcomes. These mapped probabilities then induce the input probabilities for device 2: probabilities of outcomes for device 1 are mapped into input regions for device 2. Suppose that these mapped input probabilities for device 2 make it bubbly (i.e., give small measure to each member of some partition of the input space into bubbles).

For example, for device 1 as partially illustrated by figure 9.3a and the lower part of figure 9.3b, let device 2 be a wheel of fortune with an input space represented in the top part of figure 9.3*b*. Here each of the thirty-two out-comes from device 1 would cause inputs in distinct regions in the input space for the wheel of fortune, as partially illustrated by the tiny arrows connecting outcomes from device 1 to input regions in device 2. The inputs to device 2 that are displayed in the figure are for a single bubble with four outcomes (divided by darker lines). One should imagine, though, that device 1 has many more than thirty-two distinct outcomes, mapped into small regions of device 2's input space. If device 2 were a wheel of fortune, imagine that we are interested in outcomes defined by numbered wedges, as in roulette, rather than red and black. Then the bubbles for device 2 would be regions that allowed possible inputs that could cause any of the numbered outcomes for device 2.

According to the compression theorem, this second device—a wheel of fortune, in my example—would further "compress" frequencies in the mapping to wedge-number outcomes. If this process of one bubbly causal map device feeding into the next is iterated, so that each device compresses frequencies, it could turn out that relative frequencies of outcomes were very close to the probabilities mapped from device 1 through a series of other devices. But in the illustration, the input frequency distribution for the first device was a sort of worst case: frequencies of initial conditions were not near to the probabilities of the regions in which they fell within the input space. Nevertheless, it could turn out that frequencies of outcomes could end up very close to probabilities of outcomes for the last device in the chain.

Now, the story started with an arbitrarily specified input measure for the first device, which happened to make frequencies of initial conditions non-macroperiodic. The point now is that if there is a series of potentially bubbly causal map devices linked up in the right way, all that we need is *some* starting input measure that will make each subsequent causal map device bubbly (make input sets of small measure that have conditions leading to every outcome). Any such input measure will do the job: relative to it, probabilities of outcomes for the last device in the chain will usually be close to frequencies of those outcomes. Relative to many very different step 1 input measures, input frequencies will be far from probabilities, but the chain of bubbly devices can compress frequencies to become closer to probabilities. The series of devices could iteratively map low-level events to higher-level events. The same kind of process could take place over time if a device fed its outcomes back so that they were its own initial conditions.[4] I think it's possible that something like this stepwise compression process could be what gives rise to chances in population-environment systems, perhaps beginning with very low-level phenomena such as molecular motions or global phenomena external to a population-environment system. The proposal does not depend on Lebesgue measure, Strevens's perturbation argument, or Roberts's laws of nature, yet it still allows higher-level chances to be manipulable by changes in higher-level properties such as differences in traits that affect developmental paths (§4.2.3).

9.3.4 DISCUSSION

The stepwise compression proposal may seem too good to be true. Have I smuggled in some assumptions? In fact, the requirements for stepwise compression to give rise to the stochastic behavior we find in population-environment systems are fairly demanding, even if they seem like requirements that might possibly be satisfied by a very complex system such as a population-environment system. On the other hand, a great advantage of the stepwise compression proposal is that it doesn't depend on assuming any specific sort of input measure, such as Lebesgue measure, based on natural metrics, or assuming laws of nature, or assuming the validity of an argument like Strevens's perturbation argument. Here's what's required:

1. There must be a chain of "devices," parts of the world with persistent or recurrent causal structure, where outcomes for one device result in events within the input space of the next device. (There could be a directed graph of devices where outcomes for some devices are initial conditions for multiple devices, or where more than one device can cause inputs to another device. The graph need not be acyclic: outcomes of a device could cause its subsequent inputs, as mentioned earlier. However, for the sake of simple language I'll just speak of chains.)

2. Variation in frequencies of initial conditions for the first device in the chain can be extreme and unsystematic, but the variation can't be unbounded. (Any behavior with bounded variation could in theory be compressed by a sufficiently long chain of bubbly devices linked up in the right way.[5]) However, it's plausible that even phenomena in the world with a great degree of variation do not have unbounded variation. (Consider the "pebbles on a hillside" example above. There are only so many pebbles that can fall down a hillside during any period of time.)

3. Each device must have potential bubbles—similar sets of initial conditions that contain elements leading to each of the device's outcomes.

4. Each device except the last must have an outcome space with many distinct outcomes.

5. Distinct outcomes from device i cause inputs to device $i + 1$ in such a way that each bubble in device $i + 1$ receives many inputs to most bubbles and, more specifically, many inputs for each region leading to each outcome within each bubble. These conditions are required for the application of the compression theorem.

It's the last requirement that seems the most demanding. We need not only a chain of causal map devices of particular sorts but also ones for which the outcome space from one device is hooked up to the input space for the next device in a fairly specific way. On the other hand, all that's necessary for stepwise compression to explain patterns of frequencies in a complex system is that there be *some* way of analyzing regular causal interactions into a chain (or graph) of devices that satisfies the preceding requirements. Perhaps it's not out of the question that something as complex as a population-environment system might have that kind of structure. And such a stepwise compression system doesn't have to work perfectly; it doesn't have to generate relative frequencies that are just like those that would be produced by a truly chancy system. As I explained in section 8.2, all that's needed to make sense of practice in evolutionary biology is that there be causal imprecise probabilities with values close to real-valued probabilities. Nevertheless, the assumption that such a chain of devices exists is like a similar assumption of the percolation strategy that I challenged (§9.2.3), but the former assumption is perhaps even more difficult to satisfy. On the other hand, lumpy complexity, which might easily lead to similar conditions causing very different later conditions, makes it more plausible that the requirements of stepwise compression might possibly be satisfied.

This story may still seem like magic. Jacob Rosenthal (2012, 2016) and Chloé de Canson (2022) have criticized Strevens's and my earlier proposals for basing an interpretation of probability on nothing but causal structure and actual relative frequencies. However, if a stepwise compression story *were* correct for some system, the sense in which it depended on actual relative frequencies would be so minimal that we should accept it as an account of higher-level chance, or something close to chance. All that's needed is that the variation in initial conditions remain within wide bounds, allowing the iterative, stepwise compression devices to do their job. Conditions in any world remotely like ours that could produce initial conditions for such a stepwise compression chain would not vary enough to produce high-level outcome frequencies far from corresponding mapped outcome probabilities. That should be good enough for the metaphysics of a science such as evolutionary biology, whose claims only need to be approximately correct most of the time in the context of general facts about life on our planet. In fact, this approach does not actually require us to define an interpretation of probability in terms of stepwise compression.[6] If the stepwise compression idea worked, it would explain why modeling population-environment systems as if they were causally probabilistic succeeded—whether or not an interpretation of probability was definable in similar terms. This is a way of sidestepping Rosenthal's objection (§9.2.3) to the kind of frequency-based MM-CCS interpretation of probability that I previously advocated, and stepwise compression would explain the success of empirical practices in evolutionary biology in a way that my earlier approach couldn't.

Note, finally, that the first step in the stepwise compression story is a place where single-case propensity, in the form of fundamental indeterminism from quantum mechanics, could actually play a role in evolutionary probabilities. That is, stochastic behavior due to quantum mechanical fluctuations might be part of the base-level fluctuations in behaviors that are compressed and transformed by slightly higher-level MM-CCS systems. It's important to see, though, that this does not make population-environment chances into single-case propensities. It's the chain of MM-CCS causal structures that is doing most of the work. Given stepwise compression, it doesn't matter much what the values of the quantum mechanical propensities are. If the same variation in atomic behaviors had been produced by another, deterministic process, the higher-level probabilities would turn out the same.

9.4 Conclusion

In the previous chapter, I began to outline the main ideas of measure-map complex causal structure interpretations of probability and explained what would be necessary to apply them to population-environment systems. I think there is much to recommend this idea, even though, as we saw above, its advocates have struggled with problems concerning the sources of input measures. These problems seem especially difficult for population-environment systems, but I described a potential solution using stepwise compression.

Even if the requirements for stepwise compression are in fact satisfied by population-environment systems, there is another problem for an MM-CCS account of population-environment chances that I haven't tried to address. One would like an account of why the chances of evolutionarily relevant outcomes have values close to the numerical values that research in evolutionary biology shows them to have. Given the lumpy complexity of population-environment systems, this problem is difficult to address. That is why, I believe, no other attempt to account for chances relevant to evolutionary processes has addressed it either. We have to be satisfied, instead, with an account of population-environment chances that seems as if it might allow outcomes to have the right probabilities.

9.5 Appendix: Summary of the Compression Theorem

The general point that bubbly causal structure makes outcome frequencies closer to outcome probabilities than corresponding within-bubble conditional frequencies and conditional probabilities follows from a theorem proved in Abrams (2012c). There I called it the "microconstancy inequality theorem," but here I call it the "compression theorem." The theorem says that

$$n\beta^2 S \geq |P(A) - R(A)|,$$

where n is the number of bubbles, β is the maximum input measure for any bubble, S is the maximum frequency bubble deviation—a measure-theoretic analogue of the slope of an input distribution within a bubble—and $P(A)$ and $R(A)$ are, respectively, mapped probability and relative frequency for outcome A.[7] Frequency bubble deviation, a scaled difference between probability-weighted average number of inputs for an outcome within a bubble and the same quantity for the bubble as a whole, is defined for arbitrary probability measures, not only Lebesgue measure.[8]

Since $n\beta$ is the number of bubbles times the maximum bubble probability measure, its value is at least 1, and is equal to 1 if all bubbles have the same input probability. As the maximum bubble probability β gets smaller, the quantity on the left side of the inequality gets smaller, no matter what the "slope" S of the input distribution is—no matter how steep the changes are from one part of a bubble to another. (Note that $n\beta^2 = (n\beta)\beta$ gets smaller if β gets smaller, even if n grows: β^2 shrinks faster than n increases, since $1/n$ is the smallest possible value for β.) The theorem implies that as the maximum bubble probability β decreases, the difference $|P(A) - R(A)|$ between mapped probability and relative frequency must thereby become smaller as well. This is so even when the maximum frequency bubble deviation S is large, as it would be for figure 9.3a.

Conclusion

There is no complete and unassailable guidebook for evolutionary biology, so researchers develop concepts, assumptions, models, empirical methods, and statistical techniques gradually over time, learning from and extending each other's work. Methods and assumptions that keep working—that keep returning results that seem to be more or less consistent with other things learned—will get reused. What doesn't fit might get reused too, but then there is a greater possibility of criticism, revision, and eventual abandonment. As a result, common, long-standing research practices can form, as it were, an abstract outline or silhouette of characteristics of the world that allow common practices to continue to work. In this book, I've developed a proposal about what fills part of that shape, providing a description of some of the structures in the world that could allow empirical research in evolutionary biology to function as it does.

I argued that what is studied in microevolutionary biology should be seen as based on a combination of a population and its environment that together help to make up a chance setup. This chance setup is a piece of the world that realizes objective probabilities for potential sequences of complex population and environment states. It's what I call a *population-environment system*, and I argued that population-environment systems realize *causal probabilities*, or at least *causal imprecise probabilities*, over outcomes that are also realized by those complex states. These are objective probabilities—possibly imprecise—in which manipulating probabilities by changing the causal structure of the setup usually manipulates frequencies of outcomes in a corresponding way.

I also argued that the world contains many overlapping population-environment systems, which can implicitly be selected by researchers' decisions about what populations to study and how to study them. Complex causal processes in the world don't necessarily have clear boundaries, but researchers can delineate population-environment systems that are parts of those processes and take into account, in one way or another, influences on organisms that cross the boundaries researchers have tacitly chosen. What researchers then learn about are causal processes involving organisms interacting in the piece of the world that has thus been delineated.

I focused on a handful of examples of empirical research that I believe illustrate practices that are widespread. However, even given their wide use, my focus was quite restricted. I argued for particular ways of thinking about natural selection, discussing drift and a few other evolutionary influences only briefly. While natural selection can be investigated without fitness concepts, a large part of my focus was on fitness. I proposed that fitness concepts come in four primary varieties. I argued that the sort of fitness concepts that provided the starting point for a great deal of thoughtful, illuminating discussion in philosophy of biology—*causal token-organism fitness*—is largely irrelevant to evolutionary biology. A related point is that actual organisms, despite their obvious importance, should be seen merely as the means by which population-environment systems realize traits. It's *causal organism-type fitness*—trait fitness as a summary of probable causal relations in a population-environment system—that is most closely connected to natural selection, and fitness of this kind bears only tenuous connections to causal token-organism fitness. Actual organisms are, of course, a primary focus of empirical research, as they should be. Just as we can usually learn about the bias of a pair of dice by tossing the dice many times, we can learn about causal type fitness for a trait in a population-environment system by measuring properties of actual organisms—realizers of the trait—including some organism properties sometimes described as fitnesses. I consider the latter *measurable token-organism fitnesses*; they are often summarized by what I call *statistical organism-type fitnesses* as a means of estimating causal organism-type fitnesses.

In the last chapters of the book, I explored ways of understanding what it is about population-environment systems that allows them to realize causal probabilities. What allows a population-environment system to behave in such a way that biologists are able to make sense of evolutionary processes in terms of probabilities? I rejected some obvious answers and offered a sketch of a way that nearly deterministic processes involved in population-environment systems might generate what I call *measure-map complex causal structure* probability. The basic idea of this kind of probability is not new, but the application of the approach to population-environment systems is new, and I provided a new development of a strategy to respond to what I call the input measure problem.

Despite my strategy of focusing mainly on natural selection and fitness concepts, the view that evolution takes place in population-environment systems is not simply a claim about natural selection. It is a more general claim about the nature of evolutionary processes and their study. My discussion of Weir and Cockerham's strategy for estimating F_{ST} in chapter 1 illustrates this point. Their characterization of F_{ST} as depending on "genetic sampling" from "replicate populations," I suggested, is simply a way of talking about a population-environment system that persists for many generations and produces groups worth treating as subpopulations. I mentioned that F_{ST} can be used to get evidence for or against natural selection on particular genetic loci, but that is not its only use. Other uses depend on genetic sampling and replicate populations too—that is, partly on population-environment systems, according to my view.

Among the dimensions of evolutionary processes that I discussed briefly or not at all are group-level and neighbor effects, inheritance of various sorts (genetic, nongenetic, biased, horizontal), mutation rates and biases, recombination, genetic linkage, linkage disequilibrium, nonrandom mating, differential persistence, influences from and on developmental processes, migration, and various kinds of niche construction. The idea that population-environment systems incorporate all of these influences provides, I believe, a common ground for thinking about such influences. By seeing evolutionary processes as taking place in population-environment systems rather than in collections of actual organisms, we incorporate probabilistic influences not only on actual organisms but also on possible organisms, not only on organisms as static entities but also as participating in developmental processes and life histories, not only on organisms in isolation but also in their relationships to other organisms and to changing environments, and not only on organisms individually but also as members of a population whose overall characteristics can influence its probable evolution.

Though I think that the view presented in this book provides a good answer to the questions it's intended to address, some of my conclusions in the preceding chapters may be too strong. The claims I've made may be too categorical, too determinate, for the messy reality of the world in which an evolving, complex population is embedded. If so, then what I have offered is a kind of model—a philosophical model—of evolutionary processes and their relationship to evolutionary biology. Such a model succeeds if it helps us to see the structure—a composite of ontologically real patterns—that is implicit in the world.

Even with that qualification, I wouldn't consider the view that I've presented in this book to be final. As I mentioned in the preface and other places,

I have work in progress investigating extensions, complements, and further applications of the population-environment system view, and I hope that others who see the view as valuable will consider using it as a starting point for further developments or refinements. Most of the arguments I gave depended on facts about evolutionary biology, but analogous arguments might be given, in some cases, for other areas of science. Exploration of such possibilities may be valuable.

There are no doubt flaws in the view as I have presented it, however. Criticism and proposals for revision are welcome. It may be that some aspects of my view seem quite different from previous ideas about evolution. Some aspects certainly conflict with assumptions that have provided common ground for a great deal of discussion in philosophy of biology. Thus some readers may feel that the view presented in the book is counterintuitive—possibly even radical. My goal was not to elicit such a reaction, however. I simply think that the population-environment system view provides the best way of understanding evolutionary biology and its research practices, and I hope that those who disagree with my conclusions will want to explain why the view doesn't succeed. If I've mischaracterized common practices in evolutionary biology, or left out practices that fit poorly with the population-environment system conception, I want to know. Perhaps there are theoretical ideas in evolutionary biology that don't fit well with the view, or metaphysical assumptions I've made that lead to inconsistency. It may even be that it's wrong to view evolutionary biology as a successful science in the way needed to justify my arguments. Finally, someone else may see a very different conception of evolutionary processes that fits the shape of practice and theory in evolutionary biology more closely. I hope that readers of this book, whether sympathetic or critical, will find that the ideas I've presented succeed in deepening our understanding of evolution and evolutionary biology.

Acknowledgments

There have been many influences on the long developmental process that eventually culminated in this book. I can only highlight a few people whose significance I haven't forgotten. Several of the people mentioned below deserve much greater acknowledgment, but for the sake of brevity, I'll mention only a few specific contributions. No one mentioned should be blamed for whatever may be misguided about my views.

Years ago, Ruth Millikan pushed me to think more seriously about probability and evolution when she asked, in response to one of my early discussions, "But where do the probabilities come from?" (My answer in its most recent form is in chapters 8 and 9.) My concept of causal probability arose as a response to a penetrating question at the 2010 meeting of the Philosophy of Science Association by an audience member whose name I never learned. The concept developed into its current form largely in response to Jossi Berkovitz's persistent challenges to drafts of Abrams (2015). My focus on details of empirical research in evolutionary biology was motivated and stimulated most substantially by Bill Wimsatt, as well as by Robert Brandon, David Allison, Yann Klimentidis, and the milieu of the Philosophy Department at the University of Alabama at Birmingham when it was led by Harold Kincaid.

Harold, Roberta Millstein, Josh May, and Bruce Weir gave me helpful feedback on partial or full drafts of this book. Josh also convinced me to change the title from my original choice; the present title is inspired by his suggestion. Stephen Stearns was kind enough to respond to some of my questions about his paper that I discuss in chapters 6 and 8 (Byars et al. 2010). A discussion with Elay Shech about a precursor of the population-environment conception presented in chapter 1 led to improvements. I presented material from the book at the University of Washington in 2018 (once to philosophers and once to statistical geneticists), at the Alabama Philosophical Society in 2018, and at the International Society for History, Philosophy, and Social Studies of Biology in 2019. Comments from audience members at these presentations led to improvements reflected in the book.

Conversations and feedback on some earlier papers and presentations helped to refine and sometimes substantially shape ideas that led to this book. Those to whom I am grateful include Norman Abrams, André Ariew, Murat Aydede, Prasanta Bandyopadhyay, Anouk Barberousse, Gillian Barker, Matthew Barker, John Beatty, Carl Bergstrom, Joseph Berkovitz, John Bickle, Frédéric Bouchard, Kristin Boyce, Robert Brandon, Michael Bruno, Guo-Bo Chen, Greg Cooper, Carl Craver, Lindley Darden, Lane DesAutels, Eric Desjardins, Hugh Desmond, Isabelle Drouet, Benjamin Feintzeig, Patrick Forber, Sara Franceschelli, Dan Garber, Stuart Glennan, Bruce Glymour, Nancy Hall, Jon Hodge, Carl Hoefer, Kent Holsinger, Paul Humphreys, Philippe Huneman, Yoichi Ishida, William Kallfelz, Harold Kincaid, Ehud Lamm, Francoise Longy, Joshua May, Mohan Matthen, Alexander Meehan, Francesca Merlin, Ruth Millikan, Roberta Millstein, Juan Montana, Wayne Myrvold, John Norton, Samir Okasha, Jun Otsuka, Daniel Parker, Trevor Pearce, Charles Pence, Jeremy Pober, Grant Ramsey, Thomas Reydon, Robert Richardson, Jan-Willem Romeijn, Alirio Rosales, Alex Rosenberg, Jacob Rosenthal, Eric Seidel, Elay Shech, Elliott Sober, Samuel Spinner, Kim Sterelny, Michael Strevens, Jackie Sullivan, Yoshiyuki Suzuki, Elizabeth A. Thompson, Jos Uffink, Joel Velasco, Denis Walsh, Scott Weingart, Bruce Weir, Paul Weirich, Michael Weisberg, Bill Wimsatt, Jim Woodward, and others, including anonymous reviewers of papers. I'm particularly grateful to two reviewers of the book proposal and manuscript—Rob Skipper and an anonymous individual—for their very helpful feedback.

Philippe Huneman encouraged me to write a book on evolution for many years. I finally followed his advice! Alison Chapman, Greg Pence, Alan Thompson, Joseph Wood, and Rebecca Bach provided important stimulus and help in the steps leading up to this book project. I'm very grateful to my editor at the University of Chicago Press, Karen Darling, for her patience and faith in the project, and for a number of useful suggestions. Thanks are also due for careful copyediting and indexing by Carrie Love and Theresa Wolner, respectively, on what may have been an unusually challenging manuscript.

There are deeper influences on this book. Bill Wimsatt was my adviser in graduate school, and although I'm not sure either of us realized it at the time, his influence on my thinking was and has been pervasive, and it can be seen throughout this book. More generally, I'm grateful to everyone in the University of Chicago Department of Philosophy during my time there as a student, particularly for the department's emphasis on attentive, charitable reading and interpretation of historical texts—a strategy that I direct primarily at writings of contemporary biologists and philosophers. Bill Wimsatt, Josef Stern, Dan Garber, Howard Stein, Leonard Linksy, and Manley Thompson were especially influential in this regard. Murat Aydede, David Malament, and Ron McClamrock were influential in other respects. Early in my career, I had a postdoc at the Duke Center for the Philosophy of Biology, where Robert Brandon and Alex Rosenberg taught. That year continues to have a great influence on my thinking. I also got early advice and encouragement from Kim Sterelny. Eric Schliesser provided a crucial, supportive intervention in my academic progress, without which this book would not even have become a gleam in my eye; I cannot sufficiently express my thanks for that.

I am keenly aware of ways in which my parents, Lois Barnett and Norman Abrams, helped me to develop intellectual tendencies that made this book possible. Their influence appears on every page. If I had to highlight only one aspect of each's influence, I would mention Lois's loving support for open-minded exploration of ideas and Norman's playful emphasis on rigorous investigation of reasoning and language. I'm grateful for my sister Julie's unfailing enthusiasm for all of my philosophical projects.

Finally, I'm eternally grateful for the transformation of my life afforded by sharing it with my life partner for over thirty years, Brenda Smith. It's not just that she supported my work on the project over a long period of time. I can't imagine how the book would have been written at all, or who I would be, without her.

I've borrowed and edited brief passages or figures from some of my previously published articles. These passages are reprinted with permission from copyright holders of the following:

Marshall Abrams, "Environmental Grain, Organism Fitness, and Type Fitness," in *Entangled Life: Organism and Environment in the Biological and Social Sciences*, ed. Gillian Barker, Eric Desjardins, and Trevor Pearce (Springer, 2014), 127–52.

Marshall Abrams, "Fitness 'Kinematics': Altruism, Biological Function, and Organism-Environment Histories," *Biology & Philosophy* 24, no. 4 (2009): 487–504.

Marshall Abrams, "Implications of Use of Wright's F_{ST} for the Role of Probability and Causation in Evolution," *Philosophy of Science* 79, no. 5 (December 2012): 596–608.

Marshall Abrams, "Measured, Modeled, and Causal Conceptions of Fitness," *Frontiers in Genetics* 3, no. 196 (October 2012): 1–12.

Marshall Abrams, "Mechanistic Probability," *Synthese* 187, no. 2 (2012): 343–75.

Marshall Abrams, "Mechanistic Social Probability: How Individual Choices and Varying Circumstances Produce Stable Social Patterns," in *Oxford Handbook of Philosophy of Social Science*, ed. Harold Kincaid (Oxford University Press, 2012), 184–226.

Marshall Abrams, "Populations, Pigeons, and Pomp: Prosaic Pluralism about Evolutionary Causes," *Studies in History and Philosophy of Science Part C: Studies in History and Philosophy of Biological and Biomedical Sciences* 44, no. 3 (September 2013): 294–301.

Marshall Abrams, "Probability and Manipulation: Evolution and Simulation in Applied Population Genetics," *Erkenntnis* 80, no. S3 (2015): 519–49.

Notes

Preface

1. To simplify phrasing, I often use "evolutionary biology" for the study of microevolutionary processes, ignoring other important areas of evolutionary biology.

2. My metaphorical use of "machine" (see §1.4.1) is not an allusion to philosophical debates about whether mechanism concepts (Glennan 1996; Machamer, Darden, and Craver 2000) are useful for characterizing evolutionary processes (Skipper and Millstein 2005; Barros 2008; Havstad 2011; Craver and Darden 2013; DesAutels 2016, 2017; Glennan 2017). Also note that I use "process" as an informal, common term rather than as an invocation of the view that the correct ontology of biology involves processes rather than things (Nicholson and Dupré 2018). This book might be seen as providing support for either mechanist or processualist views of evolution—notwithstanding the views' putative incompatibility (Dupré and Nicholson 2018)—but I am not committed to either.

3. "Metaphysics" refers to a category of questions and studies in rigorous philosophy (van Inwagen and Sullivan 2020; Waters 2017) that is sometimes described as being concerned with "the nature of reality." Popular uses of the term have other senses.

Introduction

1. However, I don't think explicit connections between toy models and empirical research should be a requirement for meaningful philosophy of science. Authors can contribute different elements to the collective enterprise of philosophy of science.

Chapter Zero

1. The subsets must also satisfy a technical requirement I don't describe here: they must be "measurable."

2. It's also possible to treat propositions as the arguments of the probability function.

3. In some contexts, it may be necessary to use other definitions of conditional probability (Hájek 2003; Miranda and de Cooman 2014), but this standard definition will be good enough for our purposes.

4. A random variable is not a variable and need not be random!

5. Figures 0.1 and 0.2 later in this chapter would illustrate density functions if allele frequency were a continuous quantity.

6. Some people use "chance" in narrower or broader senses. In Abrams (2012c), I used "chance" only for single-case objective probability, which I define near the end of §0.2.3. Millstein (2011) analyzed a variety of uses of "chance" in evolutionary biology and philosophy of biology. Schaffer (2007) writes of "epistemic chance," which is a kind of Bayesian probability (see also Bradley 2017; Gallow 2021).

7. See, e.g., Gillies (2000) and Hájek (2019) for more thorough surveys of interpretations of probability and their advantages and disadvantages.

8. My discussion in this section is largely independent of debates about relative virtues of Bayesian vs. frequentist statistics, which I mention in various places later in the book. Though these two schools of statistical methodology draw their initial motivation from Bayesian and frequency interpretations of probability, in practice their use need not be firmly tied to either.

9. I use "*simple* finite frequency theory" because some rather complex interpretations of probability (Strevens 2011; Abrams 2012c) can be viewed as frequency theories due to the (nonsimple) ways in which they depend on actual finite frequencies. See §9.2.3.

10. Such a set is known as a "reference class"—a class with reference to which frequencies are to be calculated.

11. This is true of Hoefer's (2007) theory despite his emphasis on what he calls "statistical nomological machines" (SNMs), such as the wheels of fortune that I describe in chapters 8 and 9. The problem is that Hoefer also defines chances in terms of frequencies alone when no such SNMs exist.

12. Propensity interpretations were originally proposed by Popper (1957), although there are earlier antecedents in the work of Peirce. Propensity interpretations of probability are not particularly related to propensity scores or any other statistical concept involving the term "propensity."

13. In biology, the type/token distinction is often made, but different terminology is used in different contexts. For example, Gillespie (2004, 25) discusses ways that it can come about that "two alleles" are "identical by state," which is to say that two token alleles share the same type in the sense of sharing the same sequence of DNA base pairs, roughly speaking.

14. Darwin introduced Herbert Spencer's phrase "the survival of the fittest" into *On the Origin of Species* in the fifth edition (1869).

15. Usually, "genotype" refers to a pattern of genes on one or more chromosomes, and that's what I mean here. "Phenotype" is usually a bodily or behavioral trait, often influenced by genotypes. "Trait" usually refers to phenotypes, but genotypes are sometimes considered traits as well. In quantitative genetics (chapter 7), a genotype is often a theoretical entity that plays a certain role in modeling and statistical inference. In practice, it's not always known that only genetic material plays the genotype role. There may be nongenetic elements that are passed to offspring through the mother's egg cells, for example. In humans, culture can be passed from parents to offspring. Each of these could turn out to be part of what is modeled as "genotype" without rendering quantitative genetic models inaccurate. Quantitative genetics is not only about genetics.

16. Byerly and Michod (1991) also intend their concept to capture intuitions about fit with the environment.

17. "Expectation" and "expected number of X" will have absolutely no *cognitive* connotations in this book. They refer to mathematical averaging operations or their results, and one should not necessarily expect (in a cognitive sense) that the expectation or expected number

of something will occur. Suppose we flip a fair coin. If the outcome is heads, I give you a dollar; if it's tails, you give me a dollar. The (mathematical) expected change to your wealth is 0, since both heads and tails have a chance of 1/2, and $1 × 1/2 + $(−1) × 1/2 = $0. However, you should not expect that result. Your wealth will change.

18. Note that one can adopt a view like Mills and Beatty's or Brandon's without assuming that the relevant probabilities are propensities per se (Drouet and Merlin 2015; see also Brandon 1990; Abrams 2007a, 2014). Also note that advocates of such views sometimes use "propensities" ambiguously both for probabilities and for the expectation defined in terms of them, though expectations are not probabilities (see Drouet and Merlin 2015).

19. Sober's views in 1984 were significantly more subtle than this summary reveals, as are his more recent views (Sober 2000, 2013, 2020). Sober (2020) suggests that Mills and Beatty (1979) did not claim that trait fitness is parasitic on fitnesses of token organisms. I disagree—see especially (Mills and Beatty 1979, 272ff.)—but the disagreement won't matter to anything important here. I'll note, however, that Mills and Beatty's other way of discussing trait fitness, in terms of optimality analyses, seems incompatible with their claim that trait fitnesses are equal to the average of token fitnesses; see §3.3.4 for some relevant discussion.

20. Fitness is often represented by "w." *Variance* is a measure of how much a probabilistically determined quantity X tends to vary. It's defined as var $X = E[(X − E(X))^2]$. This measures the difference between values that X takes (the first term on the right-hand side) and the overall average (expected) value for X (the second term). Squaring the quantity inside the outer expectation operator means that it doesn't matter whether the value of X is above or below its average value. Variance is then the average of this squared quantity. Other "central moments" can be defined by substituting higher powers for the power of 2 above. For example, skew is defined as $E[(X − E(X))^3]$, and kurtosis is defined as $E[(X − E(X))^4]$.

21. Beatty and Finsen (1989), probably the earliest philosophical paper on this topic, gave an illustration showing that the third statistical moment, skew (see note 20), must in some cases be incorporated into a fitness measure. In Abrams (2009b), I gave an example showing that the fourth moment, kurtosis, can also matter, and I pointed out that Gillespie's work implies that higher statistical moments can matter as well. See §4.5 for related discussion.

22. Some terminology: A diploid genetic system is one in which each parent has two copies of most chromosomes (as with humans). Because of ambiguities in "gene," we speak of *alleles* (variant genes) at a *locus* (a region on a given chromosome). The idea is that if we look at an organism's two copies of, say, chromosome 3, there is (typically) a particular place on the chromosome where the DNA sequence that specifies a given gene can be found. However, different tokens of chromosome 3 (either within one organism or in different organisms) may have variant DNA sequences in this region—that is, they have different alleles. A haploid genetic system is one in which each parent has only one of each kind of chromosome, so there is only one allele at each genetic locus.

23. In organisms whose cells usually contain two copies of each chromosome, gametes, such as sperm and egg cells in many vertebrates, contain only one of each chromosome.

24. A MacOS executable and scripts for generating the figures are available at https://github .com/mars0i/book/tree/master/ocaml. The OCaml source code, which can be compiled for other operating systems, is part of https://github.com/mars0i/imprecise-evolution.

25. Sometimes, as in Godfrey-Smith (2009b, 60), authors seem to treat (variance) effective population size as if it can be reified as an objective property of a population, but it's not clear to me that this assumption is defensible. Potochnik (2017, 53–54, 56, 59) views effective population

size as a way of treating a population as having a size other than the one it actually has. Whether this is correct depends on one's view of the model-world relationship.

26. "Almost": all frequencies but one would have probability 0 in the limit, but the fact that something has zero probability does not imply that it's impossible. (Because of this, biologists' use of "deterministic" as a description of such cases shouldn't be interpreted in quite the sense that this term is usually used in physics or philosophy.) For example, when tossing a coin, the probability of tossing only heads gets smaller and smaller as the number of tosses increases. This happens in such a way that the probability of only heads in an infinite sequence of tosses would be zero, but that doesn't mean that such a sequence would be impossible. In mathematical probability theory, one says that something that happens with probability 1 (so that its opposite happens with probability 0) is something that happens "almost surely," or "almost everywhere," or in "almost all" cases.

27. Biologists occasionally endorse the view that natural selection merely summarizes effects of individuals. Endler (1986) made this claim, quoting similar remarks by Ghiselin (1981). Note that "statisticalism" is called by that name not because it has to do with statistical inference. Rather, the term derives from the fact that statisticalists argue that natural selection and drift should be defined in terms of descriptive statistics that summarize facts about populations, and because they argue that population-level explanations of evolution can be given using such statistics without interpreting terms causally.

28. See, e.g., Matthen and Ariew (2002, 2009), Walsh (2007), Walsh, Ariew, and Matthen (2017), and Ariew, Rice, and Rohwer (2015); see also Ariew, Rohwer, and Rice (2017). Note that there is a burgeoning literature on noncausal explanation more generally and about relationships between mathematical explanation and science. I won't discuss this more general literature, but see, e.g., Huneman (2010, 2018), Lange (2012, 2013, 2016), Andersen (2016), Craver and Povich (2017), and Glennan (2017).

Chapter One

1. There are other contexts in evolutionary biology where replicate populations have been invoked, such as in McKane and Waxman (2007).

2. An important, related criticism (Nei and Kumar 2000, 239–40, 242) is that Cockerham and Weir's methods (Cockerham 1969, 1973; Weir and Cockerham 1984) assumed that current subpopulations were equally related to each other. Weir and colleagues have subsequently developed new F_{ST} estimation methods in order to overcome this drawback (Weir and Hill 2002; Browning and Weir 2010; Weir and Goudet 2017).

3. In July 2020, the Scopus citation index showed about 12,800 citations of Weir and Cockerham (1984). Nei (1973) was cited about 5970 times. However, there are other papers and books by both authors that could be cited as sources of F_{ST} methods.

4. Sober (1988) draws conclusions from the fact that biologists sometimes apply a statistical method, ANOVA, using properties of actual populations. The fact that Cockerham and Weir's widely used approach to estimating F_{ST} is based on extensions of ANOVA that depend on replicate populations shows that biologists' inferences need not depend only on summarizing actual facts about populations.

5. Cockerham and Weir's way of calculating F_{ST} depends on assuming that a given allele has a determinate frequency in a common ancestral population, but that frequency gets divided out and doesn't need to appear in the estimation formula.

6. Portions of this section were previously published in Abrams (2015).

7. In mathematics, outcomes are often called "events." Metaphysicians use "event" in other senses. Outcomes can be described by propositions, so many people speak of a possible outcome as a proposition.

8. To avoid awkward locutions, I'll intend "realize" in an extended sense that includes the relationships that Mellor and Suárez ($0.2.3) postulate between chance setups and the probabilities they generate.

9. The view that evolution takes place in population-environment systems is partially compatible with views like Millstein's (2006) that natural selection and random drift are population-level causal processes, but my view adds a great deal of detail about metaphysics, probability, and explicit connections to empirical practice in evolutionary biology.

10. Binomial distributions can be used to approximate Gaussian distributions (Grimmett and Stirzaker 2001). Daud (2014) argues that the distribution produced by Galton's original quincunx would not have tended to approximate a binomial distribution very well.

11. Ariew, Rohwer, and Rice (2017) discuss the Galton board in order to motivate an idea of noncausal mathematical explanation, partly inspired by some of Galton's statements. If they are right that the behavior of the Galton board can be given a noncausal explanation, that does not preclude using it as an illustration of a kind of causal process, as I do here, nor does it preclude that there are causal explanations of the same behavior (see §0.3.6).

12. See the glossary at the Internet Pinball Database (http://www.ipdb.org) for this and other pinball terms. Other helpful information came from the IPDB and responses to my questions by participants in the tiltforums.com online forum. Hands-on research helped too.

13. On magnets, a rare addition to pinball machines, see http://www.ipdb.org/glossary .php#Playfield_Magnets.

14. Interestingly, pinball machines and Galton boards may have a common origin (Pritchard 2006) in billiards-like "bagatelle" games, which required players to shoot balls on a board in which vertical pins were embedded (Ruben 2017; Natkin and Kirk 1977).

15. See Natkin and Kirk (1977, 68) and see responses to my questions at http://tiltforums .com/t/another-academic-fringe-question-highly-skilled-players-and-chance/3292/23.

16. On changes to pinball machines over time, see Ruben (2017, 28–27), Natkin and Kirk (1977, 80), and http://tiltforums.com/t/another-academic-fringe-question-highly-skilled-players -and-chance/3292.

17. See answers to my questions at http://tiltforums.com/t/academic-fringe-questions-self -playing-games-diagrams-complex-games/3278.

18. For those who think it's important to the analogy, we can even introduce quantum influences; e.g., by allowing Geiger counters to control some behaviors.

19. See Walsh (2015a) for a diametrically opposed view.

20. From UCAR Center for Science Education, see https://scied.ucar.edu/learning-zone /clouds/cloud-types and https://scied.ucar.edu/learning-zone/clouds/how-clouds-form.

21. The default conception of a population-environment system makes it a Markov process, since it can change continuously over time and the probabilities of possible states at each moment depend on the state at the previous moment (Grimmett and Stirzaker 2001).

Chapter Two

1. See, e.g., Romeijn (2017), Sprenger (2016a), Mayo (2018), Sober (2008), Royall (2004), and D. R. Anderson (2008).

2. See, e.g., Weisberg (2013), Gelfert (2016), Suárez (2009), Parker (2009), Sterrett (2014), and Odenbaugh (2021).

3. In chapter 7, I go into more detail about the pattern of statistical inference that is common to Santos et al.'s (2016a) study and many others.

4. One can view each simulation run as a model of one of the replicate populations in Cockerham and Weir's way of estimating F_{ST} (§1.1), or as a simple representation of a population sampled from among the replicate populations.

5. More specifically, the p-value is the probability that the observed relationship, or one even less like a chance effect, would have been observed if the null hypothesis were correct.

6. See also Simões, Santos, et al. (2008) and Santos et al. (2012) by some of the same authors.

7. Philosophers have distinguished "idealization" from "abstraction" and "approximation" in various ways that can matter a great deal for certain discussions (e.g., Godfrey-Smith 2009a; Norton 2012; Weisberg 2013; Morrison 2015; Appiah 2017; Potochnik 2017). Since these distinctions won't matter to my discussion, I'll lump all intended differences between what models represent and properties of the systems represented under "idealization."

8. Parts of Potochnik's (2017) argument for these points are in a chapter titled "Science Isn't after the Truth." What she means by "truth" seems to require complete accuracy of all aspects of a model or theory. However, Potochnik argues that a common goal of science is "understanding," which requires that something like PICPIE holds for a great deal of scientific research. I would rather say that science is "after the truth" in the loose sense allowed by PICPIE—which is otherwise quite consistent with Potochnik's view.

9. For example, widely used quantitative genetics methods such as those I illustrate with Byars et al. (2010) in chapter 7 had to await their development in the 1980s.

10. Illustrations of the importance of instruments in evolutionary biology include technology for analysis of genetic material, various chemical analyses relevant to physiology, light-frequency measurement of bird feather color, and waveform analysis of bird songs.

11. Potochnik's (2017) "causal pattern" derives from Dennett's (1991) "real pattern," but the latter is a pattern in observed data, whereas causal patterns are in the underlying phenomena that might generate data in a particular experimental context (see Bogen and Woodward [1988] for the data/phenomena distinction).

12. That is, typically, models in empirical research in evolutionary biology are based on "Galilean idealizations" in Weisberg's (2013) sense of an idealization made for practical reasons of tractability. However, unlike Weisberg's paradigmatic Galilean idealizations, most of these idealizations will not be removed as technology and techniques improve. Models in empirical research in evolutionary biology are not typically "minimal models," in which "only those factors that make a difference to the occurrence and essential character of the phenomenon in question" (Weisberg 2013, 100) are represented. Nor need the idealization involve what Weisberg calls "multiple models idealization," if that means that particular populations or species must be modeled in multiple ways. It's good to study a population, or a species, or similar organisms, or evolution in general, in different ways, and that will often involve models with different idealizations, but a study can provide evidence for a causal pattern without the help of other studies.

13. Portions of this section were previously published in Abrams (2015).

14. See Abrams (2015) for more on these points. Note that causal probability bears some similarities to Lyon's (2011) concept of "deterministic probability," but Lyon's goals are different from mine.

15. Berkovitz (2015) surveys versions of this strategy for propensity interpretations. A related issue and solution arise in frequentist statistical methods (Cox 2006; Mayo 2018).

16. Using C_H (or C_T) for heads (or tails), on the first flick, and E_H for heads on the second flick:

$$P(C_H|E_H) = \frac{P(E_H|C_H)P(C_H)}{P(E_H|C_H)P(C_H) + P(E_H|C_T)P(C_T)} = \frac{0.7 \times 0.6}{(0.7 \times 0.6) + (0.5 \times 0.4)} \approx 0.68.$$

The expression in the denominator is equal to $P(E_H)$.

17. Portions of this section were previously published in Abrams (2009a).

18. I'll eventually explain why the conclusions about causal probability here and in chapter 7 may need to be weakened. For the sake of a simple presentation and in keeping with the heuristic systematicity strategy mentioned in the preface, I'll put off discussing this possibility until chapter 8.

19. The simulation source code itself is fairly simple (Santos et al. 2016b), but it generates an iterative simulation process in which later states depend on earlier states in ways that are difficult to predict. Santos et al.'s code also depends on other source code such as the PRNG code provided by the R computer language.

20. Three populations are simply not enough for an investigation of relative frequencies for the processes studied. In fact, for the most part, Santos et al. (2016a) don't treat the three lab populations in aggregate; each is studied separately.

21. For evidence that using simulations to model probabilistic evolution in real populations is common, see, e.g., Hoban, Bertorelle, and Gaggiotti (2012), or simply sample the literature in evolutionary biology. For example, cursory examination of two arbitrarily chosen issues of the top journal *Evolution* (2013, volume 67, numbers 5 and 6) turned up at least six articles illustrating the same idea. One could use text-mining methods for more rigorous evidence.

22. I discuss whether and in what senses PRNGs constitute chance setups in a work in progress (Abrams PRNG MS).

23. See §6.4.4 for the applicability of this idea to coalescent models.

24. In theory, it would be preferable to provide a quantitative demonstration of assumptions about prevalence, perhaps using text-mining methods (e.g., Chavalarias et al. 2016; Ramsey and Pence 2016; Malaterre, Chartier, and Pulizzotto 2019), but I am not sure that this would even be possible with current technology.

25. This is clearly an inference to the best explanation or abduction (Douven 2017). Arguments from empirical practice are reminiscent of "no-miracles arguments" for realism (Putnam 1975), according to which the success of science would be a miracle if most of our best theories were not true—if they did not refer to real things—so we should assume that they are true. No-miracles arguments face significant challenges (Chakravartty 2017), but I don't believe that the kind of argument I am making faces the same challenges. My argument is closer in structure to Sprenger's (2016b) temporally local no-miracles argument, which does not depend on the whole history of science, but only recent history of science. In a loosely analogous way, my arguments depend only on many similar studies in relatively recent history of one science. Note also that the assumption-world correspondence that my argument requires is more limited than the full truth that some no-miracles arguments require. Indispensability arguments for realism about mathematics are also inferences to the best explanation that bear analogies to my argument, but they raise other issues (Colyvan 2019).

26. The term "*p*-hacking" comes from the fact that part of the evidence for an interesting, alternative hypothesis is a low *p*-value (§2.1); *p*-hacking can be seen as, for example, collecting data until a low *p*-value is found. Fraser et al. (2018) surveyed evolutionary biologists and ecologists, and they reported that 42% had admitted to, at a minimum, limited *p*-hacking at least once.

27. There are other issues concerning statistical testing that aren't always included under the title of questionable research practices. For example, it's been argued that results are only meaningful, or are more meaningful, if the size of effects that pass *p*-value tests are not small, and it's been argued that it's bad if the statistical "power" of a study is small. (This is the probability of getting the results one got given that the null hypothesis is false.) I'll take for granted that these are valid desiderata for research.

28. Bayesian and likelihoodist approaches to statistics may imply that a component of some sorts of *p*-hacking, namely collecting data until a desired result is obtained, can be considered scientifically legitimate in some cases (Romeijn 2017, §3.2.2–§3.2.3; Mayo 2018, §1.5; Sober 2008, §1.6). Where this is so, these critics of *p*-values can't treat this component of *p*-hacking as a questionable research practice. The issues here are complex, though.

29. This claim is based on my reading of the literature. I read in a number of areas of (micro)evolutionary biology, and the only times that I recall seeing an empirical paper that used something other than frequentist methods and *p*-values was when I performed searches for such papers. Occasionally, Bayesian methods are used in addition to frequentist methods, as in Santos et al.'s (2016a) use of a Bayesian method for estimating effective population size. My sense is that Bayesian methods are more common when phylogenetic methods are used, and they are not uncommon in some parts of multispecies ecology. It may be possible to verify claims about the prevalence of *p*-values in evolutionary biology using text-mining methods (Head et al. 2015; Chavalarias et al. 2016).

Chapter Three

1. The classification I present is almost the same as one introduced in Abrams (2012b), but some of the categories have new names, and I have refined my thinking about them.

2. Some authors, such as Sober (2020), maintain that there can be selection without heritability, in part because of models such as the breeder's equation (e.g., Rice 2004) include a kind of fitness term called a "selection differential" that's distinct from a heritability term. However, such models still allow heritable fitnesses, as applications of the models show (see chapter 7). I don't see a benefit to defining fitness concepts that allow cases in which multigenerational evolution can't take place, as the next item in the text shows.

3. This distinction between fitness roles that are tied to measurement and those that characterize causal features of evolutionary processes, which I introduced in Abrams (2012b), is close to Pence and Ramsey's (2015) later distinction between the "metrological" and "conceptual" roles of fitness. However, I don't think that what I'll call causal type fitnesses and causal token fitnesses are merely "conceptual" per se. All of the fitness categories that I discuss are defined by conceptual roles, in the sense that they are defined by how they are or can be used, but they are categories of properties of things in the world. Other taxonomies and distinctions between fitness concepts, with different purposes and scopes, can be found in Stearns (1992), de Jong (1994), Michod (1999), Reydon (2021), and other references below and in §0.3.2 (including works by Sober and by statisticalists such as Matthen, Ariew, and Walsh). Note that I don't recommend trying to map any part of my categorization onto the vernacular fitness vs. predictive fitness

dichotomy introduced by Matthen and Ariew (2002), as it carries theoretical commitments that I don't share.

4. There is much evidence that human populations continue to evolve through natural selection on genetically influenced traits. (Note, however, there seems to be no serious evidence for claims of genetically influenced, meaningful cognitive differences between racial or ethnic groups.)

5. See also, e.g., Matthen and Ariew (2002, 2009), Ariew and Ernst (2009), and Walsh (2000, 2007, 2010) for further context and evidence that statisticalists think that causalists' token fitness concepts actually refer to something real, even if the concepts may be of limited significance according to statisticalist views.

6. It would have been more accurate to call measurable token-organism fitness, for example, "relatively-easy-to-measure-from-observation-of-a-token-organism-and-its-descendants-and-nothing-else fitness." I decided against using that term despite its enhanced clarity.

7. It's possible that some statisticalists' claims about trait fitnesses make them into purely mathematical fitnesses, but I won't pursue the exegetical issue here.

8. Some arguments in this section are related to some given by Sober (1984) and Hodge (1987), who argue that overall individual fitness is not causal, but my arguments are different. Some of my arguments are also related to arguments in Ariew and Ernst (2009), but mine are more general and make it clear that it is not the propensity interpretation of fitness per se that is the problem.

9. There are discussions of similar issues in Ramsey (2006) and Pence and Ramsey (2013).

10. Wimsatt (1980b, 1981) argued that when there are nonlinear interactions between alleles, it's inappropriate to treat alleles as units of selection. Analogously, one might argue that when there are nonlinear interactions between inheritable types of any sort, it's inappropriate to assign fitness values to each type as such. However, given a probability distribution over possible combinations of types, fitness values for any one type can in principle be computed and are thus determinate. This is in effect to treat those alternative types, which might be combined with a particular type whose fitness is to be calculated, as analogous to the alternative environmental states discussed in chapter 4 (see Dawkins 1976; Sterelny and Kitcher 1988). Also see Rice (2004, 2012) for more general ways of modeling nonlinear interactions in evolution.

11. I originally gave a version of the argument in this section in Abrams (2007a) and summarized it in Abrams (2012b), but the core of the idea derives partly from Scriven (1959). Related arguments can be found in Sober (2013, 2020) and Strevens (2016). Glymour (2011; see also Glymour 2006) makes a somewhat related argument using four-way distinction between (a) individual and population environments, and (b) "narrow" and "wide" environments. A population environment is a set of conditions to which all members of the population are subject, while an individual environment is a set of conditions to which one or more organisms is subject. A narrow (individual or population) environment is specified as having specific values, while a wide (individual or population) environment is specified by a probability distribution over values. Since Glymour deals with individuals only as instances of particular traits, and since an environment in his sense is defined by a perhaps small number of variables (roughly, sets of alternative, incompatible conditions, or what some philosophers call "determinables"), a narrow individual environment is not equivalent to what I call a set of circumstances experienced by a token organism. In fact, all of Glymour's discussion concerns organism and environment type realizations or probabilities of type realization.

12. Ariew and Ernst (2009) also interpret the PIF as having this implication.

13. In Brandon's (1990, chap. 2) terms, I am arguing that when we consider environmental variation in sufficient detail, it's incorrect that there are broad regions of the space of environmental conditions that are objectively homogeneous or that vary only gradually with respect to probabilities relevant to fitness.

14. In a paper in preparation (Abrams Fitness MS), I explain why, although Pence and Ramsey's (2013) conception of fitness avoids a number of problems, it still suffers from some of the problems described in this chapter, as well as additional ones. Also note that Bouchard and Rosenberg's (2004) proposal that fitness be defined in terms of design problems may face a similar problem to the one described in this section, depending on how it is developed (see A. Rosenberg and Bouchard 2015). For example, is a particular organism solving a design problem if it's camouflaged only when it happens to sit in front of a background that is unlikely to occur more than once every ten thousand generations?

15. Sober (2013, 2020) and Ramsey (2013a, 2013b) make more subtle claims about averaging views, which I don't discuss directly. I think that other arguments in this and later chapters show that their views fail to take into account important facts about evolutionary biology.

16. I originally presented this argument in Abrams (2007a) and summarized it in Abrams (2012b). Closely related arguments can be found in Drouet and Merlin (2015), Strevens (2016), and Walsh, Ariew, and Matthen (2017).

17. See Sober (1984), Ariew and Lewontin (2004), and Walsh (2013) for related discussions.

18. Earlier, Sober (2000) used "selection for" and "selection of" without hyphens for this distinction. He first introduced a "selection for"/"selection of" distinction in Sober (1984), but the 1984 definitions differ in subtle ways from the later definitions, which are more relevant here.

19. Sober (2013) nevertheless defines type fitness as an average of actual token fitnesses (see also Sober 2001). Since type fitness so defined masks differences between neutral traits and those that make a causal difference to relative proliferation of traits, Sober distinguishes between selection-for one trait over another, on one hand, and a *fitness difference* between two traits on the other. According to Sober, that one trait B_1 is fitter than another doesn't mean that it's selected for, but only that there is at least one trait A_1 that's correlated with B_1 and that *is* selected for. Sober (2013, 340) says that it's "mistaken" to think that "variation in fitness does not cause evolution" because a "claim of fitness difference must have causal content" when "it entails a proposition that is transparently causal." I find this language a little bit puzzling, but one reading is that variation in fitness between mutually exclusive traits B_1 and B_2 can "cause" evolution even though doing so may not constitute selection for B_1 or for B_2, but only for some other trait A_1 that's correlated with one of them. This makes sense if we understand the fitness difference between B_1 and B_2 as nothing more than the difference in causal power between A_1 and A_2; that is, "variation in fitness between B_1 and B_2," is simply a way of referring to the difference in causal power between A_1 and A_2. If this is a correct reading, it still seems to me that this is an unnecessarily confusing way of expressing a significant point. See Sober (2020) for a more recent discussion of trait fitness.

20. One reason that linkage disequilibrium may persist is because of "genetic linkage"—a different concept with a related name. Genetic linkage occurs when loci are on the same chromosome and are near enough to each other that it is more likely that recombination would preserve the association between alleles among offspring, than if the loci were on different chromosomes. Note that the distinction between linkage disequilibrium and genetic linkage is pretty clear, but some authors (e.g., Gillespie 2004, §4.1) don't highlight it—presumably because the latter is a main cause of long-term persistence of the former.

21. I discuss some examples of research using linkage disequilibrium in Abrams (2015) and in a work in preparation (Abrams InfPops MS).

22. Since genetic linkage (see note 20) can be defined in terms of distances on chromosomes, it can be attributed to token organisms. However, the main reason that genetic linkage matters is because of its consequences for linkage disequilibrium.

23. If a plant's causal token fitness can be affected by other organisms in such an environment—perhaps microorganisms in the soil—then the situation is even worse for the view that plant clones provide evidence of nonextreme propensities: in that case, any organism that can affect one plant would need to be a clone of an organism that might affect any other plant, in order to preserve uniformity of circumstances (Sober 2020). But other circumstance variation might in turn affect *these* organisms as well, and thus indirectly affect plants.

24. Ramsey (2013a) also provides arguments that can be used to support the view that causal type fitnesses can't be defined in terms of a simple average of causal token fitnesses. Some of his points echo earlier points made in Abrams (2007b, 2009b), as well as an argument in Wimsatt (1972), but many of Ramsey's arguments are novel. Ramsey nevertheless argues that token fitnesses causally contribute to type fitness, but he does not spell out how. Drouet and Merlin (2015) and Bourrat (2020), too, provide arguments that can be used to motivate the idea that fitnesses of types are not parasitic on causal token fitnesses. Sober (1984; see also 2013) gave another kind of argument against the idea that there is some sort of overall causal token fitness that plays a role in evolution. This is an argument that only particular traits of a token organism can cause changes in its life, since the combination of all its traits together typically does not interact with anything, and in fact not all traits of an organism end up making a difference in its life. Other authors (Ariew and Ernst 2009; Pence and Ramsey 2015) have discussed related issues.

25. The last expression in equation (3.1) can be expanded into a sum of terms that are weighted by the probability or relative frequency P of each trait z_i:

$$E(z_i - \overline{z})(w_i - \overline{w}) = \sum_{i=1}^{n} [(z_i - \overline{z})(w_i - \overline{w})P(z_i)] =$$
$$(z_1 - \overline{z})(w_1 - \overline{w})P(z_1) + (z_2 - \overline{z})(w_2 - \overline{w})P(z_2) + \cdots + (z_n - \overline{z})(w_n - \overline{w})P(z_n).$$

Whether $P(z_i)$ refers to a probability or a relative frequency depends on how the equation is interpreted.

Chapter Four

1. Sober (2020) suggested that claims in Abrams (2009a) and Abrams (2009c) conflicted with research on selection in varying environments of the kind discussed in this chapter, but those papers made compatibility with environmental variation explicit, as did Abrams (2014), on which parts of this chapter are based.

2. Variations on this scheme can be found, e.g., in Levins (1968), Wimsatt (1980a), Gillespie (1998), Glymour (2006, 2011), and Abrams (2009a, 2014).

3. I see chapter 1 of Brandon (1990) as primarily concerned with causal token fitness. Chapter 2 alludes to causal token fitness—e.g., on page 47, when it mentions the environment of an individual. However, the primary focus of chapter 2 is on environments of populations and subpopulations of organisms, and the effects of these environments on the fitnesses of types.

4. Millstein (2014) proposed refinements of Brandon's concept of selective environment using ideas from Levins (1968). However, Millstein's presentation simplified Levins's picture by leaving out consideration of different roles of probability in Levins's ideas about fine-grained

and coarse-grained environments (cf. Wimsatt 1980a). I've tried to capture what I see as essential probabilistic features of Levins's picture in §4.2.2 and other discussions in this chapter.

5. This is a variant of the well-known statistical pattern called "Simpson's paradox." Using the first, additive model below in §4.2.2, the overall fitness $w(B)$ is greater than the overall fitness $w(A)$ if

$$w(B|E_1)P_B(E_1) + w(B|E_2)P_B(E_2) > w(A|E_1)P_A(E_1) + w(A|E_2)P_A(E_2),$$

where $P_\alpha(E_j)$ (for $\alpha = A$ or $\alpha = B$) is the probability that an organism of type α will be found in subenvironment E_j. The conditional fitness $w(\ |E_j)$ is the fitness conditional on living in environment E. (The concept of conditional fitness is common in evolutionary biology [e.g., Gillespie 1998], but often isn't given its own name, as I did in Abrams [2009a]. When fitness is expected number of offspring or some other kind of expectation, conditional fitness is a conditional expectation, a central concept of mathematical probability theory [e.g., D. Williams 1991].) For example, suppose $w(A|E_1) = 10$, $w(A|E_2) = 2$, $P_A(E_1) = 0.1$, $w(B|E_1) = 5$, $w(B|E_2) = 1$, and $P_B(E_1) = 0.9$. Then $w(A) = 10 \times 0.1 + 2 \times 0.9 = 2.8$ and $w(B) = 5 \times 0.9 + 1 \times 0.1 = 4.6$, so B is fitter than A, though A is fitter in each subenvironment. Some relevant discussion in Takacs and Bourrat's (2021) detailed criticisms of some of my papers, including Abrams (2009a), appeared too late for discussion here; explaining why my current view undermines most of their arguments would require bringing together points from many chapters in this book. I leave that job for a later work.

6. See Glymour (2011, 2014) for a more precise version of this point and Millstein (2014, esp. §4.2), for further relevant discussion.

7. Here is a specific illustration: Suppose fitness is expected number of offspring O_A for type A; i.e., $w(A) = E(O_A) = \Sigma_k k\, P(O_A = k)$. The probability of having k offspring is the average across subenvironments E_j, weighted by the probability of E_j: $P(O_A = k) = \Sigma_j P\,(O_A = k|E_j)P(E_j)$. Together these equations imply that $w(A) = E_j(w(A|E_j))$, as in the text.

8. This is known as a geometric mean. The usual mean/average/expectation (i.e., a weighted sum) is known as an arithmetic mean.

9. The way in which fitness is measured relative to each subenvironment can't be chosen arbitrarily, either, but must be tailored to the way in which fitnesses are combined. This is a lesson that I take from Glymour (2011).

10. The idea of a developmental path comes from Abrams (2009a), with earlier inspiration from ideas in Wimsatt (1972) and, I suspect, early conversations with Bill Wimsatt that reflect ideas later published in Wimsatt (2002). There are some similarities between aspects of Ramsey's (2006) concept of a fitness environment for a particular token organism and the idea of a developmental path described here, but developmental paths in my sense are defined for a population-environment system as a whole, and for types realizable in such a system. I don't see it as generally useful to view developmental paths as possible paths for a particular token organism, as Ramsey does. This is because, as remarks later in this chapter and in chapters 3 and 8 suggest, I think that what happens to a particular token organism results mainly from deterministic processes. This means that if we were to constrain a set of developmental paths to the life of a particular, realized token organism, this set would usually contain only a single path, or a collection of paths so narrowly constrained that they might as well count as a single path.

11. Ramsey (2006) suggested that the probabilities of what he called "possible lives" of organisms could be determined by counting possible lives. However, it seems that there would often be continuous variation in possible developmental paths for a given population-environment system. After all, position in space, at least, varies continuously. So the numbers of paths for a

given type would be (uncountably) infinite rather than finite. Each developmental path would usually have zero probability, while (uncountable) sets of paths could have nonzero probabilities; cf. note 26, §0.3.4. This picture then implies that the population-environment system must determine a probability density function over developmental paths (cf., e.g., Grimmett and Stirzaker 2001). Probabilities of (measurable) sets of paths would then be equal to an integral over this density.

12. See §0.3.4 for brief remarks on relationships between absolute and effective population sizes.

13. We can't just treat fitness here as an expected number of descendants, taking into account probable numbers of offspring etc. weighted by probabilities of subenvironment occurrence, as §4.2.2 and §4.2.3 show; §4.5 sketches additional reasons for this point.

14. Note that subenvironments with specific properties relevant to evolution need not have discrete boundaries. For example, there can be continuous temperature variation as one goes up a mountainside, and this continuous variation can affect fitness by degrees. One might worry that in that case, the circumstances that each organism encounters might never be encountered again; the conditions would not be recurrent. However, there are nevertheless small regions of similar conditions that would recur sufficiently often that a population could evolve in response to such conditions (Abrams 2014).

15. This idea, which I first discussed in Abrams (2007a) as a well as in later publications (Abrams 2012b, 2015), illustrates how Ramsey's (2013a) claim that type fitness depends on token fitness might be fleshed out in some cases.

16. Since concepts of propensity have played a large role in philosophical discussions of fitness, it's worth noting that in Abrams (2007a) I argued that circumstance probabilities such as the probabilities of O_iE_j mentioned above are unlikely to be propensities; see chapter 8.

17. See references in this chapter, chapter 3, and §0.3.2.

18. The title of Abrams (2009b), "The Unity of Fitness," was misleading, since the paper ultimately argued for a disunified view of fitness.

19. This view is consistent with, but may go beyond, Waters's (2017, 2019) "no general structure" view.

20. Gouvêa (2021) gives a detailed presentation of a theoretical framework for thinking about senses of simultaneous conceptual change and conceptual stasis, but I became aware of Gouvêa's work too recently to incorporate it into this discussion.

21. Characterizing Waismann's term "open texture" this way is anachronistic. Further elucidation would require a digression on the logical empiricist tradition within which his remarks occurred.

22. Hesse, who is well known for her work on the role of analogy in science (Hesse 1966), denies a clear distinction between metaphorical and nonmetaphorical language in the longer passage (Hesse 1987, 311) from which E. F. Keller (2002, 118) quotes. This view, like some similar and more recent views of metaphor (e.g., Lakoff and Johnson 2003; Gibbs and Colson 2012; Hofstadter and Sander 2013), is not without motivation. However, such views seem to require that there be no clear literal/figurative distinctions in cognitive processing, a claim I find implausible. My argument doesn't depend on this extreme assumption.

23. Consider, for example, the way in which Dawkins's (1976) "meme," referring to any relatively discrete unit of cultural inheritance, has come to refer primarily to clever images and videos distributed through social media. It seems clear from some of the examples in Hofstadter and Sander (2013) that everyday language does routinely involve creative shifts of meaning for new contexts.

24. When I suggested to an obesity researcher that he was using "thrifty gene hypothesis" incorrectly because Neel had defined it as a hypothesis about diabetes, the blank stare I received suggested that I was the one confused about meaning—a hypothesis confirmed by further reading.

Chapter Five

1. As I indicated in the preface, I won't address interesting, related questions about what can count as a group to which some form of group selection might apply.

2. One salient feature of Matthewson's (2015) population concept is based on Godfrey-Smith's (2009b) view that concepts concerning evolution, such as natural selection and drift, should be conceptualized as being a matter of degree, with some cases counting as more or less paradigmatic. Matthewson thus takes himself to be defining only what he views as the most paradigmatic sort of biological population, with departures from central cases allowed.

3. One might ask, for example, about a situation in which a female house sparrow ignores the nest that a male house sparrow built to attract females, whether this situation would count as an unsuccessful mating.

4. The researchers used bacterial strains that could not engage in lateral gene transfer; see §7.3.2.

5. This case also doesn't fit Matthewson's (2015) population definition, but he could claim that my hypothetical noninteractive *E. coli* populations simply aren't "*paradigm* Darwinian populations." Note though that since the actual experiment doesn't allow very much interaction between *E. coli*, it wouldn't have paradigm Darwinian populations either—despite the importance that has been attached to this research by biologists.

6. Millstein (2010a) also defined a concept of "metapopulation" for studies of populations whose members interact only to some degree. My remarks in the text imply that researchers might not know whether they are studying a population or metapopulation until they do further empirical research.

7. Remarks in Otsuka et al. (2011) and Gildenhuys (2014) seem to suggest that what counts as a population depends on correct methods for calculating effective population size (§2.1). Rather, it seems, given a choice of population to study, effective population size can then be defined once enough is known about the population.

8. https://www.genome.gov/10001688/international-hapmap-project.

9. http://www.cephb.fr/en/hgdp_panel.php.

10. https://www.ncbi.nlm.nih.gov/variation/news/NCBI_retiring_HapMap.

11. Moreno-Estrada et al. (2008) used a version of HGDP-CEPH that had been cleaned up according to Noah Rosenberg's (2006) specifications. Note that among other tests, Moreno-Estrada et al. use F_{ST}, described in chapter 1, and variants on the extended haplotype homozygosity tests, discussed in Abrams (2015).

12. Here the "rs" numbers refer to particular single-nucleotide polymorphisms (SNPs); i.e., chromosome locations at which the population data includes at least two different DNA base pairs. (HGDP includes data on the first two SNPs mentioned, which are near to the third SNP. That SNP is only available in the HapMap data, though.) "DAF" means "derived allele frequency," which can be an indicator of whether natural selection has affected the frequency of the SNP relative to prehuman ancestors.

13. For the record, in my brief and unsystematic survey of papers using HGDP-CEPH to investigate natural selection, most of them used some term other than "population" for continental

groups. An exception was a paper on which I had collaborated (Klimentidis et al. 2011), which analyzed HGDP-CEPH data only at the level of continental groups, calling these either "population groups" or "population." The paper was written primarily by the first author, Klimentidis, with feedback from the other authors, but apparently passed muster with reviewers.

14. That the signal—i.e., the pattern of XP-Rsb values—is less significant in the component populations is due to the fact that there is less data for each component population than for the larger population data into which their data is combined.

15. Gannett (2003) argued that the view that prevailed in the committee that defined the original, lowest-level HGDP populations assumed that demes ("populations" here) "are more-or-less discrete entities" (Gannett 2003, 997). This may be correct about the process that produced the HGDP data. However, my sense, based on reading quite a few articles that used this data and based on conversations with scientists, is that sophisticated researchers usually assume that there has probably been gene flow between populations with physically proximate histories. However, the amount of gene flow is taken to be small enough that it can be ignored in statistical analyses. I don't have the sense that researchers using HGDP data think that the division into populations by HGDP represents the only correct way of dividing up human groups.

16. In terms of the pinball analogy in chapter 1, modeling such a population is like modeling the left half of a double pinball machine in which balls from one side sometimes end up on the other side. If there is, in fact, a probability distribution over "migrations" of balls, that fact can be considered part of the left-half system. Or one can change the analogy to a narrower pinball machine in which the right wall of the playfield automatically (probabilistically) spits out balls onto the playfield.

17. To make these population concepts practical, one would need to supplement them with statistical methods for estimating quantities relevant to determining whether they apply, since determining all of the relevant causal or probabilistic relationships would be impossible in practice. Spencer (2016) discusses methods that use computer programs that apply complex statistical heuristics for picking out populations like those studied by Moreno-Estrada et al. (2009). However, those methods don't guarantee that the populations so defined satisfy philosophical definitions like the CIPC. They merely show genetic patterns that reflect past reproductive interactions on average. It might, perhaps, be argued that such methods can provide evidence of satisfaction of the PCIPC, in some cases, though I don't see the value in determining whether that is so.

18. As an illustration of population concepts, Byars et al. (2010) may be unusual, in part because it uses a detailed but preexisting data set of survey and measurement data of humans sampled from a large population in ways that were not intended for Byars et al.'s kind of study. Nevertheless, as a paper coming from a top evolutionary biologist (Stephen C. Stearns, the corresponding author) and published in a top scientific journal (*Proceedings of the National Academy of Sciences*), it's appropriate to take seriously the paper's assumptions about what counts as a population—and about which the authors use entirely common methods to infer the existence of natural selection.

19. In studies of sexually reproducing species, it's common to focus studies of natural selection on females because they are directly involved in reproduction, and because counting numbers of offspring for both males and females presents additional complications.

20. Dawber. Meadors, and Moore (1951) described the sampling methods for the FHS. In Byars et al.'s analysis, some women represented in the data set were excluded from the analysis—for example, because certain data was missing for those individuals.

21. There was presumably some unavoidable error in the collection of information on FHS participants. In addition, some information had to be left out to protect the identities of participants in the study. For example, women who had more than five children were listed as having only five children.

22. All of the properties mentioned are estimated (chapter 7), but that's not obvious from these quotations.

23. It could be that the phrasing in these quotations from Byars et al. are an artifact due to the traditional practice of applying quantitative genetic methods to things called "populations," which are often populations in some more intuitive sense. This helps to explain the language in the paper but doesn't conflict with my ultimate argument that it's possible to view the women in the study as a population.

24. Note that what's important here is not really what Byars et al. *describe as* a population but what they conceive as generating the probabilities that enter into their statistical analysis. I am using statements in their paper as evidence for this conception.

25. I try to discuss ideas from linear algebra, such as matrices and vectors, in such a way that those who are unfamiliar with them, or have forgotten them, will be able to get the gist of my points without understanding the mathematical concepts. However, in case it's helpful to have a little more structure to attach intuitions to, here is more information: Multiplying matrices (two-dimensional arrays of numbers) and/or vectors (lists of numbers) simply provides ways of multiplying many numbers and adding them up in particular, systematic ways. For example, multiplying a 3×3 matrix by a three-element (column) vector means multiplying each number in the vector by the corresponding numbers in the first row of the matrix and adding up the results, and then doing the same thing with the second and third rows. The result is a new three-element vector that might represent the result of influences represented by the numbers in the matrix on properties recorded in the vector (cf. §7.5).

26. Barker and Velasco's (2013) approach to populations is somewhat analogous to Waters's (2017, 2019) parameterized gene concept, which I mentioned in §4.5.3.

27. Matthewson (2015) incorporates a species definition due to Alan Templeton into his definition of "paradigm Darwinian population"; this allows Matthewson's concept to apply to some cases in which species boundaries are unclear. However, if one takes broader scientific concerns as relevant to what is usefully treated as a population, there's no need to build a species definition into a population concept.

Chapter Six

1. In combining "real" and "cause" here and in the title of the chapter, I'm not suggesting that I intend to engage with very interesting discussions by authors such as Hodge (1977) and Griesemer and Wade (1988) about whether natural selection, in one sense or another, has been or should be understood in terms of the Newtonian concept of *vera causa*. This is one more potentially relevant topic that would distract from my primary goals.

2. The problem of the reference environment is analogous to "the problem of the reference class" (Reichenbach 1949; Hájek 2019). On some interpretations of the latter, the former could count as an instance of it.

3. This section presents improved versions of arguments I gave in Abrams (2013). Walsh (2013) responded to that paper. He didn't directly respond to these particular arguments but did offer arguments concerning other aspects of my paper. I address those other arguments in

§6.4. (Concerning Takacs and Bourrat [2021], which also discusses Abrams [2013], see note 5 to chapter 4.)

4. We can suppose that the traits are controlled by alleles at a single locus, and that pigeons are usually homozygous for one allele or the other because the heterozygous genotype weakens the immune system. In other words, pigeons that have both a *dark* and a *light* allele suffer from more illness than other pigeons.

5. All of my claims about feral pigeons that aren't obviously made up, such as the coloring traits, are based on Johnston and Janiga (1995).

6. More precisely, assume that the distributions over numbers of descendants for *dark* in E_D is the same as that for *light* in E_L etc. Let $w(\ )$ be overall fitness of a type and $w(\ |E)$ be the fitness conditional on living in environment E (cf. note 5, §4.2.1). Thus $w(light)$ for the combined population is equal to a weighted sum of fitnesses of *light* in environments E_D and E_L, where weights depend on the proportions of *dark* and *light* in each environment. (The subpopulations are of equal size, so they contribute to overall population frequencies to the same degree). In particular, let the frequency of *light* in E_D and the frequency of *dark* in E_L be equal to p, and the frequency of each of the other two combinations be $q = 1 - p$. Then $w(light) = p\ w(light|E_D) + q\ w(light|E_L)$ and $w(dark) = q\ w(dark|E_D) + p\ w(dark|E_L)$. Since by assumption $w(light|E_D) = w(dark|E_L)$ and $w(light|E_L) = w(dark|E_D)$, it follows that $w(light) = w(dark)$. (Note that since we're assuming that fitnesses that are equal reflect identical distributions over future descendants, we can add fitnesses in this way no matter how fitness is defined over those distributions [see Abrams 2009b for relevant discussion]. Think of the preceding calculations as summing entire distributions over possible future numbers of instances of *dark* and *light*, and then apply whatever fitness function you prefer.)

7. The following arguments are based partly on Abrams (2013). See Stephens (2010), Northcott (2010), and Ramsey (2013b) for related arguments. The arguments in this section are related to some arguments concerning units of selection, in that they concern questions about whether it's appropriate to attribute causal relations to both larger and smaller groups of organisms or components of organisms (e.g., Okasha 2006). However, my argument does not depend on traditional notions of groups or interactions between or within groups. Also, Huneman's (2012, 2013, 2015) views according to which natural selection should be understood in terms of counterfactuals are potentially relevant. However, apart from what I see as the essential role of counterfactuals in frequentist statistical inference (§2.1, §7.2.3), I try to avoid making our understanding of natural selection or drift directly dependent on counterfactuals, because of problems raised by Wimsatt (1972) and Millikan (1993, 2002).

8. But see discussions of related issues in Beatty (1984) and Sober (1984, 2000, 2013, 2020).

9. The populations that Hamann, Weis, and Franks (2018) studied were originally described in Franke et al. (2006). As in Hämälä, Mattila, and Savolainen's (2018) study mentioned in §5.3.3, it's not clear precisely how these authors defined their study populations in terms of locations, which were a few miles apart.

10. This is a crude summary of Hamann, Weis, and Franks's quantitative analyses, which incorporated other responses by the plant populations as well interactions between various factors that I don't mention here.

11. This view is consistent with assumptions incorporated into models such as Weir and Cockerham's method for estimation of F_{ST} (chapter 1).

12. One can view some of Walsh's (2013) arguments as having been addressed by points I made in Abrams (2009c), but other arguments in Walsh (2013) raise entirely new challenges.

13. I'll continue to write that "Walsh would" endorse a particular conclusion, because his actual conclusions in Walsh (2013) and other works concern selection vs. drift, and I want to focus on selection alone. Although I do ultimately present arguments against the conclusions I attribute to Walsh, I have nevertheless tried to present his views fairly in order to show what I think is mistaken. Another difference between my presentation and that in Walsh (2013) is that Walsh's examples depended partly on hitchhiking due to linkage disequilibrium. The additional complexity that this introduced into Walsh's examples is not important to the points that I discuss in this section. I discussed hitchhiking in chapter 3.

14. This is similar to the sequence in Walsh's (2013) example, but the exact sequence is not essential to the points that either of us make.

15. The probability of a *particular* sequence with six *wet* and two *dry* periods (such as the one in the text) is $0.25^6 \times 0.75^2 \cong 0.00014$. For additional context, there are $\binom{8}{6} = \frac{8!}{6!2!} = 28$ such sequences, so $28 \times 0.25^6 \times 0.75^2 \cong 0.00385$ is the probability of all sequences with six *wet* and two *dry* periods. By contrast, the probability of all sequences with six *dry* and two *wet* years is $28 \times 0.75^6 \times 0.25^2 \cong 0.31146$.

16. Table 6.3 was calculated using functions in the Clojure language source file envsch/src /envsch/core.clj, available at https://github.com/mars0i/book.

17. Similar points can be applied to spatial environmental variation when there are various chances that organisms of given types will encounter particular subenvironments, and to combinations of chancy temporal and spatial variation. In future work, it might be worth engaging with an existing philosophical literature about chances whose outcomes become fixed at particular times (e.g., Lewis 1980; Lange 2006). This literature is motivated by and constrained by very different concerns than mine, so the connections are not straightforward.

18. An allele goes to fixation when the population contains no alternative alleles at the same locus. Fixation is the opposite of extinction.

19. I discussed Voight et al.'s (2006) use of coalescent simulations in Abrams (2015). Use of coalescent simulations in empirical research is quite common.

20. For the record, beginning with Abrams (2007b), my view has not been that "the conditions required for the actions of [drift or selection] are independent of the conditions required for the other" nor that "selection and drift, can be distinguished from one another by their respective effects" (Walsh 2013, 302).

21. There are also various measures of "effective population size" (§0.3.4) that allow modeling drift in the same way given various complications about interactions between organisms.

Chapter Seven

1. Suárez (2017b, 2020) discusses other relationships between chance and statistical methods in empirical research, framed partly in terms of his interpretation of probability, and Suárez (2022) applies these ideas to issues concerning fitness. I hope to discuss Suárez's views in a future publication.

2. As mentioned in §0.3.5, arguments such as those in Godfrey-Smith (2009b) led me to use the loose neologism "inheritable" in much of this book, but in this section the technical term "heritable" is correct.

3. Here "direct effects" doesn't refer to actual effects on the population; rather, it is a term contrasted with "indirect effects"—influences on the probable prevalence of a trait because of its causal relationships to other traits (Lande and Arnold 1983).

4. At another point on the same page, Byars et al. (2010) wrote, "We measured significant linear selection gradients." This use of "measured" might suggest that the selection gradients were measured directly rather than estimated; i.e., that a selection gradient is a realized collective property of the actual token organisms in the population. However, several other sentences, including the one I quoted, make it clear that regression coefficients are estimates of selection gradients, and that selection gradients are therefore not measured directly in the study (cf. Lynch and Walsh 2018). This kind of verbal (and sometimes conceptual) slippage between estimates and what is estimated is common. I take the language that distinguishes between the two, in particular between calculated regression coefficients and selection gradients, to be the more careful usage.

5. The relationship between the estimates—regression coefficients—of selection gradients is the same kind of relationship that I described in chapter 1 for Cockerham and Weir's method of estimating F_{ST}.

6. This indirect effect of selection on one trait for the change in another trait is related to generalized hitchhiking, discussed in §3.3.5.

7. We could also infer that similar processes are likely to be present in a larger population from which the FHS data was sampled.

8. As mentioned in §2.3, I weaken this conclusion in chapter 8.

9. Arguments analogous to the one above could no doubt be made in other sciences, but the details of cases matter. I don't think we should try to generalize across experimental methods to other sciences without carefully considering the details of the methods and their role in the science in question.

10. For references to other quantitative genetical examples see, e.g., Lande 1979; Lande and Arnold 1983; Endler 1986; Falconer and Mackay 1996; Roff 1997; Lynch and Walsh 1998, 2018; Gillespie 2004.

11. One might worry that Byars et al. (2010) shouldn't count as "successful" research because it didn't report whether its predictions about future years were successful. If it turned out that the predictions were too far from the truth, that would be a reason to doubt that the analysis had captured what was causally important in the population studied. However, that in itself wouldn't conflict with my conclusion that the methodology of evolutionary biology depends on the existence of causal type fitnesses that are built partly from causal probabilities. Only if analogous analyses rarely predicted results correctly would my argument be undermined—for that would be evidence that the methodology is generally misguided and doesn't tend to approximate reality.

12. In their paper, Byars et al. (2010) don't report the p-values from the linear regression whose regression coefficients estimate the selection gradients, although most of these p-values are reported in a supplemental document. Instead Byars et al. report p-values from a different regression, a Poisson regression. This is because LRS is a count variable of roughly the kind that a Poisson distribution is designed to model (Stearns, personal communication, 2018; cf. Grimmett and Stirzaker 2001; Hilbe 2014; Lynch and Walsh 2018, 708). Thus, if one only wanted to predict LRS values from women's traits, it's more sensible to model this relationship using a Poisson regression (Hilbe 2014). Since this regression more accurately captures the trait-LRS relationship than does a linear regression, p-values for the Poisson regression are better for testing whether the relationship is significant. However, selection gradients as estimated by linear regression coefficients are what matter for evolutionary change in mean values of traits (Lande and Arnold 1983; Rice 2004, chaps. 6–7), so it is the linear regression that must be the main focus of Byars et al.'s study.

13. Because of the variety of ways that samples are taken in empirical studies, justifying the claim that sampling probabilities are causal probabilities would require a large digression, but

this claim isn't essential to what follows. The claim is at least plausible to the extent that sampling is intended to capture the effects of, say, flipping coins to determine membership in a sample. Parts of my discussion in §2.3 are relevant to this point.

14. Weir's (1996; Holsinger and Weir 2009) distinction between statistical sampling and genetic sampling, mentioned in §1.1, is one version of this idea.

15. It's worth noting that in some contexts, p-values don't reflect causal probabilities in the system under study—they *only* reflect the fact that data is sampled from a larger set of events or individuals. For example, pollsters might want to estimate how all actual voters intend to vote, without, for example, investigating social processes that give rise to voters' inclinations. In such cases, the point of sampling is simply to learn about the actual population. This does not seem to be what is going on in research such as Byars et al.'s.

16. For an example of a similar practice, note that Dong, van Kleunen, and Yu (2018) used two fitness measures for individuals: the number of offspring and the total mass of all offspring.

17. It was later discovered that there was also a difference at another locus (Tenaillon et al. 2016), but this mutation is also thought to be neutral (Peng et al. 2018).

18. T. F. Cooper (2007) later performed experiments in which bacteria from an LTEE strain were modified to enable lateral gene transfer.

19. The initial single-allele difference between the two strains caused the bacteria to have different colors when spread out in a special medium, allowing researchers to measure relative sizes of the two strains.

20. Elena and Lenski (2003) wrote in their review article that "fitness reflects the propensity to leave descendants" (458). I don't know whether this use of "propensity" is related to the use of "propensity" by philosophers (§0.2.3, §0.3.2), but immediately after this remark, Elena and Lenski illustrated it with a discussion of how to measure fitnesses of populations in the same way as it is done in the LTEE.

21. Ramsey (2013b) cited McGraw and Caswell (1996) in support of his causal token fitness view.

22. McGraw and Caswell's (1996, 51) "such studies" seems only to refer back to the "more traditional ways"; there is no other antecedent to this phrase that I can see.

23. Although population genetics and quantitative genetics are somewhat distinct, they are closely related, and textbooks in each area often include at least one chapter on the other.

24. In this sense, the causal probabilities play a role somewhat analogous to laws of nature on some non-Humean views of causation.

25. For those who want to learn more about methods like Byars et al.'s (2010), there is no one source I've found that's ideal. Roff (1997, 182–83) and Rice (2004, chap. 7) provide introductory presentations. For more, one can consult the original Lande and Arnold papers (Lande 1979; Lande and Arnold 1983), or Lynch and Walsh's textbooks (Lynch and Walsh 1998, 45–47, 171–81, 188; 2018, chaps. 13, 30). What is perhaps the most widely cited textbook on quantitative genetics, Falconer and Mackay (1996), doesn't cover multivariate methods, which are central to Byars et al.'s paper.

26. A later chapter (4) of Pigliucci and Kaplan (2006) also criticizes Lande-Arnold methods, but that chapter is focused on issues that are not as directly relevant to my concerns. Discussing it would require too much of a digression.

Chapter Eight

1. Ideas about objective imprecise probability are not as well developed as ideas about subjective or epistemic imprecise probability. Discussions of objective imprecise probability include

Walley and Fine (1982), Suppes and Zanotti (1996), Strevens (2008), Fierens, Rêgo, and Fine (2009), Hájek and Smithson (2012), Hartmann (2014), Bradley (2016), Peressini (2016), Cattaneo (2017), Fenton-Glynn (2017), and Gorban (2017). (Strevens's term for imprecise probability is "quasiprobability.")

2. The relationship between imprecise probability and frequencies raises challenges that I plan to discuss in a future work (cf. Walley and Fine 1982; Cozman and Chrisman 1997; Cattaneo 2017; de Cooman and De Bock 2017; Gorban 2017).

3. *Near* 0 or 1 since an improbable quantum mechanical event might, say, cause one side of the coin to mutate into the other, or turn the coin into a small bird that flies away.

4. See Abrams (2017) for a few citations of unusual papers that argue that some cellular processes might involve quantum mechanical effects. Wimsatt (1980a), Dennett (1984), and Glymour (2001) have suggested that in some animal species, certain behaviors that are truly stochastic are favored by evolution. Glimcher (2010), Glymour (2001), and Strevens (2003) also suggest that quantum mechanical indeterminism in molecular processes in the brain might be harnessed to produce such behavior. I think it's likely that the sorts of mechanisms that these authors allude to depend only on complex deterministic effects.

5. Single-case propensity is mysterious, too, but mysteries about the nature of chance are less objectionable when they are rooted in mysteries about the fundamental nature of the physical world.

6. Beginning in §8.4 and continuing through chapter 9, I discuss the possibility that population-environment chance is a kind of MM-CCS chance. Strevens (2011) argues that his MM-CCS interpretation of probability supports single-case chances. I've previously argued that MM-CCS chances are long-run because of the fat-chance argument (Abrams 2012c). J. Rosenthal (2010, 2012) argued for a similar position for related reasons.

7. In §2.4.2, I argued that repeated studies of the same population usually don't measure it in the same state. Those are still measurements of a single trial of the same population-environment system, but they measure different—perhaps closely related—outcomes from the same chance setup.

8. The view that population-environment systems involve long-run chances also raises well-known questions about how to conceptualize roles of probabilities in frequentist statistical testing (e.g., Romeijn 2017; Mayo 2018). This issue lies beyond the scope of this book.

9. See Strevens (1998, 2003, 2005, 2008, 2013, 2015) as well as von Plato (1982, 1983), J. B. Keller (1986), Engel (1992), Diaconis (1998), Butterfield (2011), and de Canson (2020). Von Plato's book (1994) includes a historical survey of mathematical work in the twentieth century inspired by von Kries and Poincaré.

10. That is, the possible initial conditions contained in each bubble must be more similar to each other than to those in other bubbles on each dimension that defines the points in the input space. There need not be an overall similarity relation that summarizes all of the dimensions. In Abrams (2012c, 2012d), I didn't require that possible initial conditions that make up a bubble be more similar to each other than to conditions in other sets, and this requirement isn't essential. However, it greatly simplifies presentation and seems reasonable for population-environment systems. Note that the way in which other authors such as Jacob Rosenthal and Michael Strevens define what I call bubbles, in terms of metrics defined by physical quantities (see §8.4.2), implies that bubbles satisfy the similarity requirement. (The "bubble" terminology reflects the fact that, like a soap bubble in the air, bubbles in the input spaces of causal map devices "reflect"—i.e., contain points leading to—the entire "surroundings"—i.e., the outcome space—"in miniature"—i.e., within a small region of the input space.)

11. "Macroperiodic" is Strevens's (2003) term. My usage is perhaps slightly broader than his.

12. Friction changes the reasoning only slightly (Engel 1992; Strevens 2003, 2015; Gallow 2021).

13. The kind of similarity needed for the definition of bubbles above doesn't depend on degrees of similarity—which might involve a metric—but only on an order or even a partial order, so that we can say that some elements are more similar to each other than they are to others.

14. $[u_0, u_1)$ is the set of real numbers starting from u_0, including u_0 up to u_1 but not including u_1. This "half-open" interval is preferable here to satisfy a minor mathematical requirement.

15. Lebesgue measure generalizes summed length, area, or volume to an arbitrary number of dimensions, while avoiding some technical problems by adding a few mathematical refinements that won't concern us here.

16. Strevens applies his MM-CCS account to organism types "given a smooth distribution over initial conditions" (Strevens 2016, 166) at an earlier time. These seem to be initial conditions for a process that would include at least a large part of what I call a population-environment system.

Chapter Nine

1. An earlier literature (von Plato 1994) on simple physical devices derived from Poincaré's (1912) work, and including authors such as Hopf (1934, 1935) and Hostinsky (1926), is valuable but is for the most part not directly relevant to the present project because of the particular ways in which these authors' arguments depend on limits.

2. My thinking about stepwise compression was stimulated by remarks in Strevens (2013), but I develop the ideas in my own way; see also Strevens (2016). The points I make about stepwise compression are closely related to ones that Myrvold (2021) makes about his "parabola gadget" model, but that is a device that feeds outcomes back into its input space. Myrvold nevertheless ultimately concludes that an input measure must be viewed as a theoretical Bayesian probability measure, an assumption I don't make.

3. For example, for dice tossing, the basic outcomes would be two, three, . . . , twelve.

4. For idealized examples of such devices in related contexts, see Strevens (2003) (the "autophagous wheels"), Strevens (2016), and Myrvold (2021) (the "parabola device").

5. The boundedness requirement is a weak analogue of Poincaré's (1912, §91) requirement that a croupier's distribution be absolutely continuous with respect to Lebesgue measure.

6. Strevens (2003, 2005, 2008, 2013, 2016) has repeatedly advocated explaining frequencies in terms of MM-CCS ideas without defining an interpretation of probability.

7. In Abrams (2012c) β was called π. I proved the theorem only for two outcomes, but one can easily repeat applications to handle a finite number of additional basic outcomes. Similar theorems, such as Strevens's (2003), that allow characterizing departures from ideal macroperiodicity only informally (as well as being restricted to Lebesgue measure) don't support the conclusion needed here as clearly as the compression theorem does.

8. Specifically, where N is the total number of inputs in an initial condition distribution, E_A (this was E_a in Abrams [2012c]) is the probability-weighted average number of inputs for outcome A (i.e., heights of boxes in figure 9.2), and E_b is the probability-weighted average number of inputs for bubble b (horizontal dashed lines in figure 9.2), the frequency bubble deviation for outcome A in b is $|E_b - E_A|/(NP(B))$.

References

Abrams, Marshall. 2007a. "Fitness and Propensity's Annulment?" *Biology and Philosophy* 22 (1): 115–30.

———. 2007b. "How Do Natural Selection and Random Drift Interact?" *Philosophy of Science* 74 (5): 666–79.

———. 2009a. "Fitness 'Kinematics': Altruism, Biological Function, and Organism-Environment Histories." *Biology & Philosophy* 24 (4): 487–504.

———. 2009b. "The Unity of Fitness." *Philosophy of Science* 76 (5): 750–61.

———. 2009c. "What Determines Biological Fitness? The Problem of the Reference Environment." *Synthese* 166 (1): 21–40.

———. 2012a. "Implications of Use of Wright's F_{ST} for the Role of Probability and Causation in Evolution." *Philosophy of Science* 79 (5): 596–608.

———. 2012b. "Measured, Modeled, and Causal Conceptions of Fitness." *Frontiers in Genetics* 3 (196): 1–12.

———. 2012c. "Mechanistic Probability." *Synthese* 187 (2): 343–75.

———. 2012d. "Mechanistic Social Probability: How Individual Choices and Varying Circumstances Produce Stable Social Patterns." In *Oxford Handbook of Philosophy of Social Science*, edited by Harold Kincaid, 184–226. Oxford University Press.

———. 2013. "Populations, Pigeons, and Pomp: Prosaic Pluralism about Evolutionary Causes." *Studies in History and Philosophy of Science Part C: Studies in History and Philosophy of Biological and Biomedical Sciences* 44 (3): 294–301.

———. 2014. "Environmental Grain, Organism Fitness, and Type Fitness." In *Entangled Life: Organism and Environment in the Biological and Social Sciences*, edited by Gillian A. Barker, Eric Desjardins, and Trevor Pearce, 127–52. Springer.

———. 2015. "Probability and Manipulation: Evolution and Simulation in Applied Population Genetics." *Erkenntnis* 80 (S3): 519–49.

———. 2017. "Probability and Chance in Mechanisms." In *The Routledge Handbook of Mechanisms and Mechanical Philosophy*, edited by Stuart Glennan and Phyllis Illari, 169–83. Routledge.

———. 2019. "Natural Selection with Objective Imprecise Probability." *Proceedings of Machine Learning Research*, edited by Jasper De Bock, Cassio P. de Campos, Gert de Cooman, Erik Quaeghebeur, and Gregory Wheeler, 103:2–13.

———. Drift MS. "Dimensions of Random Drift." Unpublished MS.

———. Fitness MS. "Biological Fitness and Complexity." Unpublished MS.

———. InfPops MS. " 'Infinite Populations' in Evolutionary Models." Unpublished MS.

———. LongRun MS. "Long-Run Chance and Diverse Trials." Unpublished MS.

———. PRNG MS. "Pseudorandomness in Simulations and Nature." Unpublished MS.

Andersen, Holly. 2016. "Complements, Not Competitors: Causal and Mathematical Explanations." *The British Journal for the Philosophy of Science* 69 (2): 485–508.

Anderson, David R. 2008. *Model Based Inference in the Life Sciences.* Springer.

Anderson, Ted R. 2006. *Biology of the Ubiquitous House Sparrow.* Oxford University Press.

Antonovics, J., K. Clay, and J. Schmitt. 1987. "The Measurement of Small-Scale Environmental Heterogeneity Using Clonal Transplants of *Anthoxanthum odoratum* and *Danthonia spicata." Oecologia* 71 (4): 601–7.

Appiah, Kwame Anthony. 2017. *As If: Idealization and Ideals.* Harvard University Press.

Arai, Kenichi, Takahisa Harayama, Satoshi Sunada, and Peter Davis. 2012. "Randomness in a Galton Board from the Viewpoint of Predictability: Sensitivity and Statistical Bias of Output States." *Physical Review E* 86:056216.

Ariew, André, and Zachary Ernst. 2009. "What Fitness Can't Be." *Erkenntnis* 71 (3): 289–301.

Ariew, André, and Richard C. Lewontin. 2004. "The Confusions of Fitness." *British Journal for the Philosophy of Science* 55:347–63.

Ariew, André, Collin Rice, and Yasha Rohwer. 2015. "Autonomous-Statistical Explanations and Natural Selection." *The British Journal for the Philosophy of Science* 66 (3): 635–58.

Ariew, André, Yasha Rohwer, and Collin Rice. 2017. "Galton, Reversion and the Quincunx: The Rise of Statistical Explanation." *Studies in History and Philosophy of Science Part C: Studies in History and Philosophy of Biological and Biomedical Sciences* 66:63–72.

Augustin, Thomas, Frank P. A. Coolen, Gert de Cooman, and Matthias C. M. Troffaes, eds. 2014. *Introduction to Imprecise Probabilities.* Wiley.

Barker, Matthew J., and Joel D. Velasco. 2013. "Deep Conventionalism about Evolutionary Groups." *Philosophy of Science* 80 (5): 971–82.

Barrett, Rowan D. H., Stefan Laurent, Ricardo Mallarino, Susanne P. Pfeifer, Charles C. Y. Xu, Matthieu Foll, Kazumasa Wakamatsu, Jonathan S. Duke-Cohan, Jeffrey D. Jensen, and Hopi E. Hoekstra. 2019. "Linking a Mutation to Survival in Wild Mice." *Science* 363 (6426): 499–504.

Barros, D. Benjamin. 2008. "Natural Selection as a Mechanism." *Philosophy of Science* 75 (3): 306–22.

Batterman, Robert W. 2002. *The Devil in the Details: Asymptotic Reasoning in Explanation, Reduction, and Emergence.* Oxford University Press.

Beatty, John. 1984. "Chance and Natural Selection." *Philosophy of Science* 51:183–211.

Beatty, John, and Susan Finsen. 1989. "Rethinking the Propensity Interpretation: A Peek Inside Pandora's Box." In *What the Philosophy of Biology Is,* edited by Michael Ruse, 17–30. Kluwer Academic Publishers.

Beckerman, A. P., M. Boots, and K. J. Gaston. 2007. "Urban Bird Declines and the Fear of Cats." *Animal Conservation* 10 (3): 320–25.

Beisbart, Claus. 2016. "A Humean Guide to Spielraum Probabilities." *Journal of General Philosophy of Science* 47 (1): 189–216.

Berkovitz, Joseph. 2015. "The Propensity Interpretation of Probability: A Re-evaluation." *Erkenntnis* 80 (S3): 629–711.

Berthier, Pierre, Mark A. Beaumont, Jean-Marie Cornuet, and Gordon Luikart. 2002. "Likelihood-Based Estimation of the Effective Population Size Using Temporal Changes in Allele Frequencies: A Genealogical Approach." *Genetics* 160 (2): 741–51.

Bickle, John. 2018. "From Microscopes to Optogenetics: Ian Hacking Vindicated." *Philosophy of Science* 85 (5): 1065–77.

Bik, Elisabeth M., Arturo Casadevall, and Ferric C. Fang. 2016. "The Prevalence of Inappropriate Image Duplication in Biomedical Research Publications." *mBio* 7 (3).

Bogen, James, and James Woodward. 1988. "Saving the Phenomena." *The Philosophical Review* 97 (3): 303–52.

Bouchard, Frédéric. 2008. "Causal Processes, Fitness and the Differential Persistence of Lineages." *Philosophy of Science*, 560–70.

———. 2011. "Darwinism without Populations: A More Inclusive Understanding of the 'Survival of the Fittest.'" *Studies in History and Philosophy of Science Part C: Studies in History and Philosophy of Biological and Biomedical Sciences* 42 (1): 106–14.

Bouchard, Frédéric, and Alexander Rosenberg. 2004. "Fitness, Probability, and the Principles of Natural Selection." *British Journal for the Philosophy of Science* 55 (4): 693–712.

Bourrat, Pierrick. 2020. "Natural Selection and the Reference Grain Problem." *Studies in History and Philosophy of Science Part A* 80 (April): 1–8.

———. 2021. *Facts, Conventions, and the Levels of Selection*. Cambridge University Press.

Bradley, Seamus. 2016. "Vague Chance?" *Ergo* 3 (18–21): 524–38.

———. 2017. "Are Objective Chances Compatible with Determinism?" *Philosophy Compass* 12 (8): e12430.

Brainard, Lindsay. 2020. How to Explain How-Possibly. *Philosophers' Imprint* 20 (13): 1–23. http://hdl.handle.net/2027/spo.3521354.0020.013.

Brandon, Robert N. 1978. "Adaptation and Evolutionary Theory." *Studies in the History and Philosophy of Science* 9 (3): 181–206.

———. 1990. *Adaptation and Environment*. Princeton University Press.

———. 2005. "The Difference between Selection and Drift: A Reply to Millstein." *Biology and Philosophy* 20:153–70.

Brandon, Robert N., and Janis Antonovics. 1996. "The Coevolution of Organism and Environment." In *Concepts and Methods in Evolutionary Biology*, 161–78. Cambridge University Press.

Brandon, Robert N., and John Beatty. 1984. "The Propensity Interpretation of 'Fitness'—No Interpretation Is No Substitute." *Philosophy of Science* 51:342–47.

Brandon, Robert N., and Scott Carson. 1996. "The Indeterministic Character of Evolutionary Theory: No 'No Hidden Variables Proof' but No Room for Determinism Either." *Philosophy of Science* 63:315–37.

Brigandt, Ingo, and Alan C. Love. 2012. "Conceptualizing Evolutionary Novelty: Moving beyond Definitional Debates." *Journal of Experimental Zoology Part B: Molecular and Developmental Evolution* 318 (6): 417–27.

Browning, Sharon R., and Bruce S. Weir. 2010. "Population Structure with Localized Haplotype Clusters." *Genetics* 185 (4): 1337–44.

Bulmer, M. G. (1967) 1979. *Principles of Statistics*. 2nd ed. Dover.

Burian, Richard M. 1983. "Adaptation." In *Dimensions of Darwinism*, edited by Marjorie Grene, 286–314. Cambridge University Press.

Butterfield, J. 2011. "Less Is Different: Emergence and Reduction Reconciled." *Foundations of Physics* 41 (6): 1065–135.

Byars, Sean G., Douglas Ewbank, Diddahally R. Govindaraju, and Stephen C. Stearns. 2010. "Natural Selection in a Contemporary Human Population." *Proceedings of the National Academy of Sciences* 107 (S1): 1787–92.

Byerly, Henry C., and Richard E. Michod. 1991. "Fitness and Evolutionary Explanation." *Biology & Philosophy* 6:1–22.

Caraco, Thomas. 1981. "Risk-Sensitivity and Foraging Groups." *Ecology* 62 (3): 527–31.

Caswell, Hal. 2001. *Matrix Population Models.* Sinauer.

Catoni, C., A. Peters, and H. M. Schaefer. 2009. "Dietary Flavonoids Enhance Conspicuousness of a Melanin-Based Trait in Male Blackcaps but Not of the Female Homologous Trait or of Sexually Monochromatic Traits." *Journal of Evolutionary Biology* 22:1649–57.

Cattaneo, Marco E. G. V. 2017. "Empirical Interpretation of Imprecise Probabilities." In *Proceedings of the Tenth International Symposium on Imprecise Probability: Theories and Applications,* edited by Alessandro Antonucci, Giorgio Corani, Inés Couso, and Sébastien Destercke, PMLR 62:61–72.

Chakravartty, Anjan. 2017. "Scientific Realism." In *The Stanford Encyclopedia of Philosophy* (Summer 2017), edited by Edward N. Zalta. https://plato.stanford.edu/archives/sum2017/entries/scientific-realism/.

Charlesworth, Brian, and Deborah Charlesworth. 2010. *Elements of Evolutionary Genetics.* Roberts and Company.

Chavalarias, David, Joshua David Wallach, Alvin Ho Ting Li, and John P. A. Ioannidis. 2016. "Evolution of Reporting P Values in the Biomedical Literature, 1990–2015." *JAMA* 315 (11): 1141–48.

Clatterbuck, Hayley. 2015. "Drift Beyond Wright-Fisher." *Synthese* 192 (11): 3487–507.

Clatterbuck, Hayley, Elliott Sober, and Richard C. Lewontin. 2013. "Selection Never Dominates Drift (nor Vice Versa)." *Biology & Philosophy* 28 (4): 577–92.

Cockerham, C. Clark. 1969. "Variance of Gene Frequencies." *Evolution* 23:72–84.

———. 1973. "Analyses of Gene Frequencies." *Genetics* 74:679–700.

Colyvan, Mark. 2019. "Indispensability Arguments in the Philosophy of Mathematics." In *The Stanford Encyclopedia of Philosophy* (Spring 2019), edited by Edward N. Zalta. https://plato.stanford.edu/archives/spr2019/entries/mathphil-indis/.

Cooper, Tim F. 2007. "Recombination Speeds Adaptation by Reducing Competition between Beneficial Mutations in Populations of *Escherichia coli.*" *PLOS Biology* 5 (9): 1–7.

Cooper, Tim F., Daniel E. Rozen, and Richard E. Lenski. 2003. "Parallel Changes in Gene Expression After 20,000 Generations of Evolution in *Escherichia coli.*" *Proceedings of the National Academy of Sciences* 100 (3): 1072–77.

Cooper, William S. 1984. "Expected Time to Extinction and the Concept of Fundamental Fitness." *Journal of Theoretical Biology* 107:603–29.

———. 2001. *The Evolution of Reason.* Cambridge University Press.

Cox, David R. 2006. *Principles of Statistical Inference.* Cambridge University Press.

Cozman, Fabio, and Lonnie Chrisman. 1997. "Learning Convex Sets of Probability from Data." Tech. Report, CMU-RI-TR-97-25. Robotics Institute, Carnegie Mellon University.

Craver, Carl F. 2022. "Toward an Epistemology of Intervention: Optogenetics and Maker's Knowledge." In *The Tools of Neuroscience Experiment: Philosophical and Scientific Perspectives,* edited by John Bickle, Carl F. Craver, and Ann-Sophie Barwich. Routledge.

Craver, Carl F., and Lindley Darden. 2013. *In Search of Mechanisms: Discoveries across the Life Sciences.* University of Chicago Press.

Craver, Carl F., and Mark Povich. 2017. "The Directionality of Distinctively Mathematical Explanations." *Studies in History and Philosophy of Science Part A* 63:31–38.

Crozier, Gillian. 2010. "A Formal Investigation of Cultural Selection Theory: Acoustic Adaptation in Bird Song." *Biology and Philosophy* 25 (5): 781–801.

Darwin, Charles. (1859) 1964. *On the Origin of Species by Means of Natural Selection, or the Preservation of Favoured Races in the Struggle for Life.* Facsimile of the first edition, with an introduction by Ernst Mayr. Harvard University Press.

———. 1869. *On the Origin of Species by Means of Natural Selection, or the Preservation of Favoured Races in the Struggle for Life.* 5th ed. John Murray. http://darwin-online.org.uk/content /frameset?ItemID=F0070&viewtype=text&pageseq=1.

Daud, Auni Aslah Mat. 2014. "Mathematical Modelling and Symbolic Dynamics Analysis of Three New Galton Board Models." *Communications in Nonlinear Science and Numerical Simulation* 19 (10): 3476–91.

Dawber, Thomas R., Gilcin F. Meadors, and Felix E. Moore Jr. 1951. "Epidemiological Approaches to Heart Disease: The Framingham Study." *American Journal of Public Health* 41:279–86.

Dawkins, Richard. 1976. *The Selfish Gene.* 1st ed. Oxford University Press.

de Canson, Chloé. 2022. "Objectivity and the Method of Arbitrary Functions." *The British Journal for the Philosophy of Science.* Published online August 2022. DOI: 10.1093/bjps/axaa001.

de Cooman, Gert, and Jasper De Bock. 2017. "Computable Randomness is Inherently Imprecise." In *Proceedings of the Tenth International Symposium on Imprecise Probability: Theories and Applications*, edited by Alessandro Antonucci, Giorgio Corani, Inés Couso, and Sébastien Destercke, PMLR 62:133–144.

de Jong, G. 1994. "The Fitness of Fitness Concepts and the Description of Natural Selection." *Quarterly Review of Biology* 69 (1): 3–29.

Dennett, Daniel C. 1984. *Elbow Room: The Varieties of Free Will Worth Wanting.* MIT Press.

———. 1991. "Real Patterns." *Journal of Philosophy* 88 (1): 27–51.

DesAutels, Lane. 2016. "Natural Selection and Mechanistic Regularity." *Studies in History and Philosophy of Science Part C: Studies in History and Philosophy of Biological and Biomedical Sciences* 57:13–23.

———. 2017. "Mechanisms in Evolutionary Biology." In *The Routledge Handbook of Mechanisms and Mechanical Philosophy*, edited by Stuart Glennan and Phyllis Illari, 296–307. Routledge.

de Valpine, Perry. 2000. "A New Demographic Function Maximized by Life-History Evolution." *Proceedings of the Royal Society of London. Series B, Biological Sciences* 267:357–62.

Diaconis, Persi. 1998. "A Place for Philosophy? The Rise of Modeling in Statistical Science." *Quarterly of Applied Mathematics* 56:797–805.

Dobzhansky, Theodosius. 1970. *Genetics of the Evolutionary Process.* Columbia University Press.

Dong, Bi-Cheng, Mark van Kleunen, and Fei-Hai Yu. 2018. "Context-Dependent Parental Effects on Clonal Offspring Performance." *Frontiers in Plant Science* 9:1824.

Dor, Roi, and Arnon Lotem. 2009. "Heritability of Nestling Begging Intensity in the House Sparrow (*Passer domesticus*)." *Evolution* 63 (3): 738–48.

Douven, Igor. 2017. "Abduction." In *The Stanford Encyclopedia of Philosophy* (Summer 2017), edited by Edward N. Zalta. https://plato.stanford.edu/archives/sum2017/entries/abduction/.

Drouet, Isabelle, and Francesca Merlin. 2015. "The Propensity Interpretation of Fitness and the Propensity Interpretation of Probability." *Erkenntnis* 80 (S3): 457–68.

Dupré, John, and Daniel J. Nicholson. 2018. "Manifesto for a Processual Philosophy of Biology." In *Everything Flows: Towards a Processual Philosophy of Biology*, edited by Daniel J. Nicholson and John Dupré, 3–45. Oxford University Press.

Eagle, Antony. 2004. "Twenty-One Arguments against Propensity Analyses of Probability." *Erkenntnis* 60 (3): 371–416.

Earman, John. 1992. *Bayes or Bust*. MIT Press.

Earnshaw, Eugene. 2018. *Modelling Evolution: A New Dynamical Account*. Routledge.

Elena, Santiago F., and Richard E. Lenski. 2003. "Evolution Experiments with Microorganisms: The Dynamics and Genetic Bases of Adaptation." *Nature Reviews Genetics* 4:457–69.

Endler, John A. 1986. *Natural Selection in the Wild*. Princeton University Press.

Engel, Eduardo M. R. A. 1992. *A Road to Randomness in Physical Systems*. Springer-Verlag.

Ereshefsky, Marc. 1992. "Eliminative Pluralism." *Philosophy of Science* 59 (4): 671–90.

Evans, Karl L., Kevin J. Gaston, Stuart P. Sharp, Andrew McGowan, and Ben J. Hatchwell. 2009. "The Effect of Urbanisation on Avian Morphology and Latitudinal Gradients in Body Size." *Oikos* 118:251–59.

Ewens, Warren J. 2004. *Mathematical Population Genetics 1: Theoretical Introduction*. 2nd ed. Springer.

Falconer, Douglas S., and Trudy F. C. Mackay. 1996. *Introduction to Quantitative Genetics*. 4th ed. Prentice Hall.

Fanelli, Daniele. 2009. "How Many Scientists Fabricate and Falsify Research? A Systematic Review and Meta-Analysis of Survey Data." *PLOS ONE* 4 (5): 1–11.

Fenton-Glynn, Luke. 2017. "Imprecise Best System Chances." In *EPSA15 Selected Papers*, edited by Michela Massimi, Jan-Willem Romeijn, and Gerhard Schurz, 297–308. Vol. 5 of *European Studies in Philosophy of Science*. Springer.

Fierens, Pablo I., Leandro Chaves Rêgo, and Terrence L. Fine. 2009. "A Frequentist Understanding of Sets of Measures." *Journal of Statistical Planning and Inference* 139:1879–92.

Franke, Denise M., Allan G. Ellis, Manisha Dharjwa, Melinda Freshwater, Miki Fujikawa, Alejandra Padron, and Arthur E. Weis. 2006. "A Steep Cline in Flowering Time for *Brassica rapa* in Southern California: Population-Level Variation in the Field and the Greenhouse." *International Journal of Plant Sciences* 167 (1): 83–92.

Fraser, Hannah, Tim Parker, Shinichi Nakagawa, Ashley Barnett, and Fiona Fidler. 2018. "Questionable Research Practices in Ecology and Evolution." *PLOS ONE* 13 (7): 1–16.

Friedman, Nathaniel A. 1970. *Introduction to Ergodic Theory*. Van Nostrand Reinhold.

Gallow, J. Dimitri. 2021. "A Subjectivist's Guide to Deterministic Chance." *Synthese* 198 (5): 4339–72.

Galván, Ismael. 2018. "Predation Risk Determines Pigmentation Phenotype in Nuthatches by Melanin-Related Gene Expression Effects." *Journal of Evolutionary Biology* 31 (12): 1760–71.

Gannett, Lisa. 2003. "Making Populations: Bounding Genes in Space and in Time." *Philosophy of Science* 70 (5): 989–1001.

Gardner, M., A. González-Neira, O. Lao, F. Calafell, J. Bertranpetit, and D. Comas. 2005. "Extreme Population Differences across Neuregulin 1 Gene, with Implications for Association Studies." *Molecular Psychiatry* 11 (September): 66–75.

Gelfert, Axel. 2016. *How to Do Science with Models: A Philosophical Primer*. Springer.

Gelman, Andrew, John B. Carlin, Hal S. Stern, David B. Dunson, Aki Vehtari, and Donald B. Rubin. 2013. *Bayesian Data Analysis*. 3rd ed. CRC Press, Taylor; Francis Group.

Gelman, Andrew, and Deborah Nolan. 2002. "You Can Load a Die, but You Can't Bias a Coin." *The American Statistician* 56 (4): 308–11.

Ghiselin, Michael T. 1981. "Categories, Life, and Thinking." *Behavioral and Brain Sciences* 4 (2): 269–283.

Gibbs, Raymond W., Jr., and Herbert L. Colson. 2012. *Interpreting Figurative Meaning*. Cambridge University Press.

Giere, Ronald N. 1973. "Objective Single-Case Probabilities and the Foundations of Statistics." In *Logic, Methodology and Philosophy of Science IV*, edited by Patrick Suppes, Leon Henkin, Athanase Joja, and Gr. C. Moisil, 467–83. North-Holland.

Gildenhuys, Peter. 2009. "An Explication of the Causal Dimension of Drift." *British Journal for the Philosophy of Science* 60:521–55.

———. 2014. "Arbitrariness and Causation in Classical Population Genetics." *The British Journal for the Philosophy of Science* 65 (3): 429–44.

Gillespie, John H. 1973. "Polymorphism in Random Environments." *Theoretical Population Biology* 4 (2): 193–95.

———. 1974. "Natural Selection for Within-Generation Variance in Offspring Number." *Genetics* 76:601–6.

———. 1975. "Natural Selection for Within-Generation Variance in Offspring Number: II. Discrete Haploid Models." *Genetics* 81:403–13.

———. 1977. "Natural Selection for Variances in Offspring Numbers: A New Evolutionary Principle." *American Naturalist* 111:1010–14.

———. 1998. *Population Genetics: A Concise Guide*. Johns Hopkins University Press.

———. 2004. *Population Genetics: A Concise Guide*. 2nd ed. Johns Hopkins University Press.

Gillies, Donald A. 2000. *Philosophical Theories of Probability*. Routledge.

Glennan, Stuart. 1996. "Mechanisms and the Nature of Causation." *Erkenntnis* 44:49–71.

———. 2017. *The New Mechanical Philosophy*. Oxford University Press.

Glimcher, Paul W. 2010. *Foundations of Neuroeconomic Analysis*. Oxford University Press.

Glymour, Bruce. 2001. "Selection, Indeterminism, and Evolutionary Theory." *Philosophy of Science* 68 (4): 518–35.

———. 2006. "Wayward Modeling: Population Genetics and Natural Selection." *Philosophy of Science* 73:369–89.

———. 2011. "Modeling Environments: Interactive Causation and Adaptations to Environmental Conditions." *Philosophy of Science* 78 (3): 448–71.

———. 2014. "Adaptation, Adaptation to, and Interactive Causes." In *Entangled Life: Organism and Environment in the Biological and Social Sciences*, edited by Trevor Pearce, Gillian A. Barker, and Eric Desjardins, 105–26. Springer.

Godfrey-Smith, Peter. 2007. "Conditions for Evolution by Natural Selection." *Journal of Philosophy* 104:489–516.

———. 2009a. "Abstractions, Idealizations, and Evolutionary Biology." In *Mapping the Future of Biology: Evolving Concepts and Theories*, edited by Anouk Barberousse, Michel Morange, and Thomas Pradeu, 47–56. Boston Studies in the Philosophy of Science. Springer.

———. 2009b. *Darwinian Populations and Natural Selection*. Oxford University Press.

Gorban, Igor I. 2017. *The Statistical Stability Phenomenon*. Springer International Publishing AG.

Gouvêa, Devin Susannne Yagel. 2021. "Essentially Dynamic Concepts and the Case of 'Homology.'" PhD thesis, University of Chicago. Committee on Conceptual and Historical Studies of Science.

Graves, Leslie, Barbara L. Horan, and Alex Rosenberg. 1999. "Is Indeterminism the Source of the Statistical Character of Evolutionary Theory?" *Philosophy of Science* 66:140–57.

Gremer, Jennifer R., Elizabeth E. Crone, and Peter Lesica. 2012. "Are Dormant Plants Hedging Their Bets? Demographic Consequences of Prolonged Dormancy in Variable Environments." *The American Naturalist* 179 (3): 315–27.

Griesemer, James R., and Michael J. Wade. 1988. "Laboratory Models, Causal Explanation and Group Selection." *Biology and Philosophy* 3 (1): 67–96.

Griffiths, Paul, and Karola Stotz. 2013. *Genetics and Philosophy: An Introduction.* Cambridge University Press.

———. 2014. "Conceptual Barriers to Interdisciplinary Communication." In *Enhancing Communication & Collaboration in Interdisciplinary Research*, edited by Michael O'Rourke, Stephen Crowley, Sanford D. Eigenbrode, and J. D. Wulfhorst, 195–215. Sage Publications, Inc.

Grim, Patrick, Robert Rosenberger, Adam Rosenfeld, Brian Anderson, and Robb E. Eason. 2013. "How simulations fail." *Synthese* 190:2367–2390.

Grimmett, G. R., and D. R. Stirzaker. 2001. *Probability and Random Processes.* 3rd ed. Oxford University Press.

Haasl, Ryan J., and Bret A. Payseur. 2016. "Fifteen Years of Genomewide Scans for Selection: Trends, Lessons and Unaddressed Genetic Sources of Complication." *Molecular Ecology* 25 (1): 5–23.

Hacking, Ian. 1965. *The Logic of Statistical Inference.* Cambridge University Press.

Hájek, Alan. 1996. "'Mises Redux'—Redux: Fifteen Arguments against Finite Frequentism." *Erkenntnis* 45 (2–3): 209–27.

———. 2003. "What Conditional Probability Could Not Be." *Synthese* 137:273–323.

———. 2009. "Fifteen Arguments against Hypothetical Frequentism." *Erkenntnis* 70 (2): 211–35.

———. 2019. "Interpretations of Probability." In *The Stanford Encyclopedia of Philosophy* (Winter 2012), edited by Edward N. Zalta. https://plato.stanford.edu/archives/fall2019/entries/probability-interpret.

Hájek, Alan, and Michael Smithson. 2012. "Rationality and Indeterminate Probabilities." *Synthese* 187 (1): 33–48.

Hämälä, Tuomas, Tiina M. Mattila, and Outi Savolainen. 2018. "Local Adaptation and Ecological Differentiation under Selection, Migration, and Drift in *Arabidopsis lyrata.*" *Evolution* 72 (7): 1373–86.

Hamann, Elena, Arthur E. Weis, and Steven J. Franks. 2018. "Two Decades of Evolutionary Changes in *Brassica rapa* in Response to Fluctuations in Precipitation and Severe Drought." *Evolution* 72 (12): 2682–96.

Hartmann, Stephan. 2014. "Imprecise Probabilities in Quantum Mechanics." In *Foundations and Methods from Mathematics to Neuroscience*, edited by Colleen E. Crangle, Aldofo García de la Sienra, and Helen Longino, 77–82. CSLI Publications.

Havstad, Joyce C. 2011. "Problems for Natural Selection as a Mechanism." *Philosophy of Science* 78 (3): 512–23.

Head, Megan L., Luke Holman, Rob Lanfear, Andrew T. Kahn, and Michael D. Jennions. 2015. "The Extent and Consequences of *p*-Hacking in Science." *PLOS Biology* 13 (3): 1–15.

Hein, Jotun, Mikkel H. Schierup, and Carsten Wiuf. 2005. *Gene Genealogies, Variation and Evolution: A Primer in Coalescent Theory.* Oxford University Press.

Hesse, Mary. 1987. "Tropical Talk: The Myth of the Literal." *Proceedings of the Aristotelian Society, Supplementary Volumes* 61:297–311.

Hesse, Mary B. 1966. *Models and Analogies in Science.* University of Notre Dame Press.

Hilbe, Joseph M. 2014. *Modeling Count Data.* Cambridge University Press.

Hoban, Sean, Giorgio Bertorelle, and Oscar E. Gaggiotti. 2012. "Computer Simulations: Tools for Population and Evolutionary Genetics." *Nature Reviews Genetics* 13 (2): 110–22.

Hodge, M. J. S. 1977. "The Structure and Strategy of Darwin's 'Long Argument.'" *The British Journal for the History of Science* 10 (3): 237–246.

———. 1987. "Natural Selection as a Causal, Empirical, and Probabilistic Theory." In *The Probabilistic Revolution*, edited by Lorenz Krüger, Gerd Gigerenzer, and Mary S. Morgan, 2:233–70. MIT Press.

Hoefer, Carl. 2007. "The Third Way on Objective Probability: A Sceptic's Guide to Objective Chance." *Mind* 116 (463): 449–596.

———. 2019. *Chance in the World: A Humean Guide to Objective Chance*. Oxford University Press.

Hoffmann, Ary A., and Loren H. Rieseberg. 2008. "Revisiting the Impact of Inversions in Evolution: From Population Genetic Markers to Drivers of Adaptive Shifts and Speciation?" *Annual Review of Ecology, Evolution, and Systematics* 39:21–42.

Hofstadter, Douglas R., and Emmanuel Sander. 2013. *Surfaces and Essences: Analogy as the Fuel and Fire of Thinking*. Basic Books.

Holsinger, Kent E., and Bruce S. Weir. 2009. "Genetics in Geographically Structure Populations: Defining, Estimating and Interpreting F_{ST}." *Nature Reviews Genetics* 10:639–50.

Hopf, Eberhard. 1934. "On Causality, Statistics and Probability." *Journal of Mathematics and Physics* 13:51–102.

———. 1935. "Remarks on Causality and Probability." *Journal of Mathematics and Physics* 14:4–9.

Hostinsky, Bohuslav. 1926. "Sur la méthode des fonctions arbitraires dans le calcul des probabilités." *Acta Mathematica* 49:95–113.

Howson, Colin, and Peter Urbach. 1993. *Scientific Reasoning: The Bayesian Approach*. 2nd ed. Open Court.

Hu, Yang, and Gonçalo C. Cardoso. 2009. "Are Bird Species That Vocalize at Higher Frequencies Preadapted to Inhabit Noisy Urban Areas?" *Behavioral Ecology* 20 (6): 1268–73.

Humphreys, Paul. 1985. "Why Propensities Cannot Be Probabilities." *Philosophical Review* 94 (4): 557–70.

Huneman, Philippe. 2010. "Topological Explanations and Robustness in Biological Sciences." *Synthese* 177 (2): 213–45.

———. 2012. "Natural Selection: A Case for the Counterfactual Approach." *Erkenntnis* 76 (2): 171–94.

———. 2013. "Assessing Statistical Views of Natural Selection: Room for Non-local Causation?" *Studies in History and Philosophy of Science Part C: Studies in History and Philosophy of Biological and Biomedical Sciences* 44 (4, Part A): 604–12.

———. 2014. "Mapping an Expanding Territory: Computer Simulations in Evolutionary Biology." *History and Philosophy of the Life Sciences* 36 (1): 60–81.

———. 2015. "Inscrutability and the Opacity of Natural Selection and Random Genetic Drift: Distinguishing the Epistemic and Metaphysical Aspects." *Erkenntnis* 80 (S3): 491–518.

———. 2018. "Outlines of a Theory of Structural Explanations." *Philosophical Studies* 175 (3): 665–702.

Ioannidis, John P. A. 2005. "Why Most Published Research Findings Are False." *PLOS Medicine* 2 (8): 0696–0701.

Ishida, Yoichi. 2009. "Sewall Wright and Gustave Malécot on Isolation by Distance." *Philosophy of Science* 76:784–96.

Johnston, Richard E., and Marian Janiga. 1995. *Feral Pigeons*. Oxford University Press.

Keller, Evelyn Fox. 2002. *Making Sense of Life: Explaining Biological Development with Models, Metaphors, and Machines*. Harvard University Press.

Keller, Joseph B. 1986. "The Probability of Heads." *The American Mathematical Monthly* 93 (3): 191–97.

Kirkpatrick, Mark. 2010. "How and Why Chromosome Inversions Evolve." *PLOS Biology* 8 (9): 1–5.

Kirschel, Alexander N. G., Daniel T. Blumstein, Rachel E. Cohen, Wolfgang Buermann, Thomas B. Smith, and Hans and Slabbekoornc. 2009. "Birdsong Tuned to the Environment: Green Hylia Song Varies with Elevation, Tree Cover, and Noise." *Behavioral Ecology* 20 (5): 1089–95.

Klimentidis, Yann C., Marshall Abrams, Jelai Wang, Jose R. Fernandez, and David B. Allison. 2011. "Natural Selection at Genomic Regions Associated with Obesity and Type-2 Diabetes: East Asians and Sub-Saharan Africans Exhibit High Levels of Differentiation at Type-2 Diabetes Regions." *Human Genetics* 129 (4): 407–18.

Kneusel, Ronald T. 2018. *Random Numbers and Computers*. Springer.

Knuth, Donald E. 1998. *The Art of Computer Programming*. Vol. 2, *Seminumerical Algorithms*. 3rd ed. Addison-Wesley.

Kozlov, V. V., and M. Yu. Mitrofanova. 2003. "Galton Board." *Regular and Chaotic Dynamics* 8 (4): 431–439.

Kunert, Joachim, Astrid Montag, and Sigrid Pölmann. 2001. "The Quincunx: History and Mathematics." *Statistical Papers* 42:143–69.

Kuriyan, John, Boyana Konforti, and David Wemmer. 2013. *The Molecules of Life: Physical and Chemical Principles*. 1st ed. Garland Science.

Lakoff, George, and Mark Johnson. 2003. *Metaphors We Live By*. 2nd ed. University of Chicago Press.

Lande, Russell. 1979. "Quantitative Genetic Analysis of Multivariate Evolution, Applied to Brain:Body Size Allometry." *Evolution* 33 (1): 402–16.

Lande, Russell, and Stevan J. Arnold. 1983. "The Measurement of Selection on Correlated Characters." *Evolution* 37 (6): 1210–26.

Lange, Marc. 2006. "Do Chances Receive Equal Treatment under the Laws? Or: Must Chances Be Probabilities?" *The British Journal for the Philosophy of Science* 57 (2): 383–403.

———. 2012. "What Makes a Scientific Explanation Distinctively Mathematical?" *The British Journal for the Philosophy of Science* 64 (3): 485–511.

———. 2013. "Really Statistical Explanations and Genetic Drift." *Philosophy of Science* 80 (2): 169–88.

———. 2016. *Because without Cause: Non-causal Explanations in Science and Mathematics*. Oxford University Press.

Laplane, Lucie, Paolo Mantovani, Ralph Adolphs, Hasok Chang, Alberto Mantovani, Margaret McFall-Ngai, Carlo Rovelli, Elliott Sober, and Thomas Pradeu. 2019. "Opinion: Why Science Needs Philosophy." *Proceedings of the National Academy of Sciences* 116 (10): 3948–52.

L'Ecuyer, Pierre, and Richard Simard. 2007. "TestU01: A C Library for Empirical Testing of Random Number Generators." *ACM Transactions on Mathematical Software* 33 (4): 22.

Lenski, Richard E. 2017. "Experimental Evolution and the Dynamics of Adaptation and Genome Evolution in Microbial Populations." *The ISME Journal* 11:2181–94.

Lenski, Richard E., Michael R. Rose, Suzanne C. Simpson, and Scott C. Tadler. 1991. "Long-Term Experimental Evolution in *Escherichia coli*. I. Adaptation and Divergence During 2,000 Generations." *The American Naturalist* 138 (6): 1315–41.

Lenski, Richard E., Michael J. Wiser, Noah Ribeck, Zachary D. Blount, Joshua R. Nahum, J. Jeffrey Morris, Luis Zaman, et al. 2015. "Sustained Fitness Gains and Variability in Fitness Trajectories in the Long-Term Evolution Experiment with *Escherichia coli.*" *Proceedings of the Royal Society of London B: Biological Sciences* 282:20152292.

Levi, Isaac. 1980. *The Enterprise of Knowledge.* MIT Press.

Levins, Richard. 1968. *Evolution in Changing Environments.* Princeton University Press.

Lewis, David. 1973. "Causation." *Journal of Philosophy*, 70 (17): 556–67.

———. 1980. "A Subjectivist's Guide to Objective Chance." In *Studies in Inductive Logic and Probability*, edited by Richard C. Jeffrey, 263–93. Vol. II. University of California Press. (Reprinted in Lewis 1986.)

———. 1986. *Philosophical Papers.* Vol. 2. Oxford University Press.

———. 1994. "Humean Supervenience Debugged." *Mind* 103:473–90.

Lewontin, Richard C. 1970. "The Units of Selection." *Annual Review of Ecology and Systematics* 1:1–18.

———. 2000. *The Triple Helix: Gene, Organism, and Environment.* Harvard University Press.

Li, Na, and Matthew Stephens. 2003. "Modeling Linkage Disequilibrium and Identifying Recombination Hotspots Using Single-Nucleotide Polymorphism Data." *Genetics* 165 (4): 2213–33.

Liu, Xuanyao, Rick Twee-Hee Ong, Esakimuthu Nisha Pillai, Abier M. Elzein, Kerrin S. Small, Taane G. Clark, Dominic P. Kwiatkowski, and Yik-Ying Teo. 2013. "Detecting and Characterizing Genomic Signatures of Positive Selection in Global Populations." *The American Journal of Human Genetics* 92 (6): 866–81.

Loewer, Barry. 2001. "Determinism and Chance." *Studies in the History and Philosophy of Modern Physics* 32B (4): 609–20.

———. 2004. "David Lewis's Humean Theory of Objective Chance." *Philosophy of Science* 71 (5): 1115–25.

———. 2020. "The Mentaculus Vision." In *Statistical Mechanics and Scientific Explanation: Determinism, Indeterminism and Laws of Nature*, edited by Valia Allori, 3–29. World Scientific.

Lorenz, Edward N. 1993. *The Essence of Chaos.* University of Washington Press.

Lue, Arthur, and Howard Brenner. 1993. "Phase Flow and Statistical Structure of Galton-Board Systems." *Physical Review E* 47:3128–44.

Lynch, Michael, and Bruce Walsh. 1998. *Genetics and Analysis of Quantitative Traits.* Sinauer.

———. 2018. *Evolution and Selection of Quantitative Traits.* Oxford University Press.

Lyon, Aidan. 2011. "Deterministic Probability: Neither Chance nor Credence." *Synthese* 182 (3): 413–32.

———. 2014. "From Kolmogorov, to Popper, to Renyi: There's No Escaping Humphreys' Paradox (When Generalized)." In *Chance and Temporal Asymmetry*, edited by Alastair Wilson, 112–25. Oxford University Press.

Machamer, Peter, Lindley Darden, and Carl F. Craver. 2000. "Thinking about Mechanisms." *Philosophy of Science* 67 (I): 25.

Malaterre, Christophe, Jean-François Chartier, and Davide Pulizzotto. 2019. "What Is This Thing Called Philosophy of Science? A Computational Topic-Modeling Perspective 1934–2015." *HOPOS: The Journal of the International Society for the History of Philosophy of Science* 9 (2): 215–49.

Malécot, Gustave. 1969. *The Mathematics of Heredity.* W. H. Freeman.

Matthen, Mohan. 2009. "Drift and 'Statistically Abstractive Explanation.'" *Philosophy of Science* 76:464–87.

Matthen, Mohan, and André Ariew. 2002. "Two Ways of Thinking about Fitness and Natural Selection." *Journal of Philosophy* 99 (2): 55–83.

———. 2009. "Selection and Causation." *Philosophy of Science* 76 (2): 201–24.

Matthewson, John. 2015. "Defining Paradigm Darwinian Populations." *Philosophy of Science* 82 (2): 178–97.

Mayo, Deborah G. 2018. *Scientific Inference as Severe Testing: How to Get Beyond the Statistics Wars*. Cambridge University Press.

McGraw, James B., and Hal Caswell. 1996. "Estimation of Individual Fitness from Life-History Data." *The American Naturalist* 147 (1): 47–64.

McKane, Alan J., and David Waxman. 2007. "Singular Solutions of the Diffusion Equation of Population Genetics." *Journal of Theoretical Biology* 247 (4): 849–58.

McLaughlin, Richard N., and Harmit S. Malik. 2017. "Genetic Conflicts: The Usual Suspects and Beyond." *Journal of Experimental Biology* 220 (1): 6–17.

Mellor, D. H. 1971. *The Matter of Chance*. Cambridge University Press.

Metz, J. A. J. 2008. "Fitness." In *Encyclopedia of Ecology*, edited by Sven Erik Jørgensen and Brian D. Fath, 2:1599–1612. Elsevier.

Michod, Richard E. 1999. *Darwinian Dynamics*. Princeton University Press.

Millikan, Ruth Garrett. 1993. "Propensities, Exaptations, and the Brain." In *White Queen Psychology and Other Essays for Alice*, 31–50. MIT Press.

———. 2002. "Biofunctions: Two Paradigms." In *Functions: New Readings in the Philosophy of Psychology and Biology*, edited by André Ariew, Robert Cummins, and Mark Perlman. Oxford University Press.

Mills, Susan, and John Beatty. 1979. "The Propensity Interpretation of Fitness." *Philosophy of Science* 46 (2): 263–86.

Millstein, Roberta L. 2002. "Are Random Drift and Natural Selection Conceptually Distinct?" *Biology and Philosophy* 17:35–53.

———. 2003. "Interpretations of Probability in Evolutionary Theory." *Philosophy of Science* 70:1317–28.

———. 2006. "Natural Selection as a Population-Level Causal Process." *British Journal for the Philosophy of Science* 57 (4): 627–53.

———. 2009. "Populations as Individuals." *Biological Theory* 4 (3): 267–73.

———. 2010a. "The Concepts of Population and Metapopulation in Evolutionary Biology and Ecology." In *Evolution Since Darwin: The First 150 Years*, edited by M. A. Bell, D. J. Futuyma, W. F. Eanes, and J. S. Levinton, 61–86. Sinauer.

———. 2010b. "Should We Be Population Pluralists? A Reply to Stegenga." *Biological Theory* 5 (3): 271–75.

———. 2011. "Chances and Causes in Biology: How Many Chances Become One Chance." In *Causality in the Sciences*, edited by Phyllis McKay Illari, Federica Russo, and Jon Williamson, 425–444. Oxford University Press.

———. 2014. "How the Concept of Population Resolves Concepts of Environment." *Philosophy of Science* 81 (5): 741–55.

———. 2015. "Thinking about Populations and Races in Time." *Studies in History and Philosophy of Science Part C: Studies in History and Philosophy of Biological and Biomedical Sciences* 52:5–11.

———. 2016. "Probability in Biology: The Case of Fitness." In *The Oxford Handbook of Probability and Philosophy*, edited by Alan Hájek and Christopher R. Hitchcock, 601–22. Oxford University Press.

Miranda, Enrique, and Gert de Cooman. 2014. "Structural Judgements." In *Introduction to Imprecise Probabilities*, edited by Thomas Augustin, Frank P. A. Coolen, Gert de Cooman, and Matthias C. M. Troffaes, 56–78. Wiley.

Mitchell, Sandra D. 2009. *Unsimple Truths: Science, Complexity, and Policy*. University of Chicago Press.

Mitchell-Olds, Thomas, and Ruth G. Shaw. 1987. "Regression Analysis of Natural Selection: Statistical Inference and Biological Interpretation." *Evolution* 41 (6): 1149–61.

Moreno-Estrada, Andrés, Ferran Casals, Anna Ramírez-Soriano, Baldo Oliva, Francesc Calafell, Jaume Bertranpetit, and Elena Bosch. 2008. "Signatures of Selection in the Human Olfactory Receptor OR5I1 Gene." *Molecular Biology and Evolution* 25 (1): 144–54.

Moreno-Estrada, Andrés, Kun Tang, Martin Sikora, Tomàs Marquès-Bonet, Ferran Casals, Arcadi Navarro, Francesc Calafell, Jaume Bertranpetit, Mark Stoneking, and Elena Bosch. 2009. "Interrogating 11 Fast-Evolving Genes for Signatures of Recent Positive Selection in Worldwide Human Populations." *Molecular Biology and Evolution* 26 (10): 2285–97.

Morris, J. Jeffrey, Richard E. Lenski, and Erik R. Zinser. 2012. "The Black Queen Hypothesis: Evolution of Dependencies through Adaptive Gene Loss." *mBio* 3 (2): 1–7.

Morrison, Margaret. 2015. *Reconstructing Reality: Models, Mathematics, and Simulations*. Oxford University Press.

Myrvold, Wayne C. 2012. "Deterministic Laws and Epistemic Chances." In *Probability in Physics*, edited by Meir Hemmo and Yemima ben Menahem, 73–85. Springer.

———. 2021. *Beyond Chance and Credence*. Oxford University Press.

Nadkarni, M. G. 1998. *Basic Ergodic Theory*. 2nd ed. Birkhäuser Verlag.

Nagylaki, Thomas. 1992. *Introduction to Theoretical Population Genetics*. Springer-Verlag.

Natarajan, Chandrasekhar, Noriko Inoguchi, Roy E. Weber, Angela Fago, Hideaki Moriyama, and Jay F. Storz. 2013. "Epistasis among Adaptive Mutations in Deer Mouse Hemoglobin." *Science* 340 (6138): 1324–27.

Natkin, Bobbye Claire, and Steve Kirk. 1977. *All about Pinball: A Fascinating Look at the Thrills, Skills, and Nostalgia of Today's Most Popular Game*. Grosset & Dunlap.

Neel, James V. 1962. "Diabetes Mellitus: A 'Thrifty' Genotype Rendered Detrimental by 'Progress'?" *American Journal of Human Genetics* 14 (4): 353–62.

Nei, Masatoshi. 1973. "Analysis of Gene Diversity in Subdivided Populations." *Proceedings of the National Academy of Sciences USA* 70 (12): 3321–23.

———. 1977. "F-Statistics and Analysis of Gene Diversity in Subdivided Populations." *Annals of Human Genetics* 41 (2): 225–33.

———. 1987. *Molecular Evolutionary Genetics*. Columbia University Press.

Nei, Masatoshi, and R. K. Chesser. 1983. "Estimation of Fixation Indices and Gene Diversities." *Annals of Human Genetics* 47 (3): 253–59.

Nei, Masatoshi, and Sudhir Kumar. 2000. *Molecular Evolution and Phylogenetics*. Oxford University Press.

Nersessian, J. Nancy. 2008. *Creating Scientific Concepts*. MIT Press.

Nesse, Randolph M., and George C. Williams. 1994. *Why We Get Sick: The New Science of Darwinian Medicine*. Random House.

Nicholson, Daniel J., and John Dupré, eds. 2018. *Everything Flows: Towards a Processual Philosophy of Biology*. Oxford University Press.

Northcott, Robert. 2010. "Walsh on Causes and Evolution." *Philosophy of Science* 77 (3): 457–67.

Norton, John D. 2012. "Approximation and Idealization: Why the Difference Matters." *Philosophy of Science* 79: 207–32.

Nuzzo, Regina. 2014. "Statistical Errors." *Nature* 506:150–52.

Odenbaugh, Jay. 2021. "Models, Models, Models: A Deflationary View." *Synthese* 198 (S21): S5061–S5076.

Odling-Smee, F. John, Kevin N. Laland, and Marcus W. Feldman. 2003. *Niche Construction: The Neglected Process in Evolution*. Princeton University Press.

Okasha, Samir. 2006. *Evolution and the Levels of Selection*. Oxford University Press.

Olson, Valérie A., and Ian P. F. Owens. 1998. "Costly Sexual Signals: Are Carotenoids Rare, Risky or Required?" *Trends in Ecology & Evolution* 13 (12): 510–14.

O'Malley, Maureen A., and Emily C. Parke. 2018. "Microbes, Mathematics, and Models." *Studies in History and Philosophy of Science Part A* 72: 1–10.

Open Science Collaboration. 2015. "Estimating the Reproducibility of Psychological Science." *Science* 349 (6251): aac4716.

Ordás, Bernardo, Rosa A. Malvar, and William G. Hill. 2008. "Genetic Variation and Quantitative Trait Loci Associated with Developmental Stability and the Environmental Correlation between Traits in Maize." *Genetics Research* 90 (5): 385–95.

Otsuka, Jun, Trin Turner, Colin Allen, and Elisabeth A. Lloyd. 2011. "Why the Causal View of Fitness Survives." *Philosophy of Science* 78 (2): 209–24.

Parker, Wendy S. 2009. "Confirmation and Adequacy-for-Purpose in Climate Modelling." *Proceedings of the Aristotelian Society, Supplementary Volumes* 83:233–49.

Pence, Charles H. 2017. "Is Genetic Drift a Force?" *Synthese* 194 (6): 1967–88.

———. 2021. *The Causal Structure of Natural Selection*. Cambridge University Press.

Pence, Charles H., and Grant Ramsey. 2013. "A New Foundation for the Propensity Interpretation of Fitness." *The British Journal for the Philosophy of Science* 64:851–81.

———. 2015. "Is Organismic Fitness at the Basis of Evolutionary Theory?" *Philosophy of Science* 82 (5): 1081–91.

Peng, Fen, Scott Widmann, Andrea Wünsche, Kristina Duan, Katherine A. Donovan, Renwick C. J. Dobson, Richard E Lenski, and Tim F. Cooper. 2018. "Effects of Beneficial Mutations in *pykF* Gene Vary over Time and across Replicate Populations in a Long-Term Experiment with Bacteria." *Molecular Biology and Evolution* 35 (1): 202–10.

Peressini, Anthony F. 2016. "Imprecise Probability and Chance." *Erkenntnis* 81 (3): 561–86.

Philippi, Tom, and Jon Seger. 1989. "Hedging One's Evolutionary Bets, Revisited." *Trends in Ecology & Evolution* 4 (2): 41–44.

Pigliucci, Massimo, and Jonathan Kaplan. 2006. *Making Sense of Evolution: The Conceptual Foundations of Evolutionary Biology*. University of Chicago Press.

Plutynski, Anya. 2007. "Drift: A Historical and Conceptual Overview." *Biological Theory* 2 (2): 156–67.

Poincaré, Henri. (1907) 1968. *La science et l'hypothèse*. Flammarion.

———. (1908) 1999. *Science et méthode*. Editions Kimé.

———. 1912. *Calcul des probabilités*. 2nd ed. Gauthier-Villars.

Popper, Karl R. 1957. "The Propensity Interpretation of the Calculus of Probability, and the Quantum Theory." In *Observation and Interpretation*, edited by S. Körner, 65–70. Academic Press Inc.

———. 1959. "The Propensity Interpretation of Probability." *British Journal for the Philosophy of Science* 10:25–42.

———. 1983. *Realism and the Aim of Science*. Routledge.

———. 1990. *A World of Propensities.* Thoemmes.

Potochnik, Angela. 2017. *Idealization and the Aims of Science.* University of Chicago Press.

Prasad, Anisha, Melanie J. F. Croydon-Sugarman, Rosalind L. Murray, and Asher D. Cutter. 2011. "Temperature-Dependent Fecundity Associates with Latitude in *Caenorhabditis briggsae.*" *Evolution* 65 (1): 52–63.

Price, George R. 1970. "Selection and Covariance." *Nature* 227:520–21.

———. 1972. "Extension of Covariance Selection Mathematics." *Annals of Human Genetics* 35: 485–590.

Pritchard, Chris. 2006. "Bagatelle as the Inspiration for Galton's Quincunx." *BSHM Bulletin: Journal of the British Society for the History of Mathematics* 21 (2): 102–10.

Putnam, Hilary. 1975. *Mathematics, Matter, and Method: Philosophical Papers.* Vol. 1. Cambridge University Press.

Ramsey, Grant. 2006. "Block Fitness." *Studies in History and Philosophy of Biological and Biomedical Sciences* 37 (3): 484–98.

———. 2013a. "Can Fitness Differences Be a Cause of Evolution?" *Philosophy and Theory in Biology* 5 (1): 1–13.

———. 2013b. "Organisms, Traits, and Population Subdivisions: Two Arguments against the Causal Conception of Fitness?" *The British Journal for the Philosophy of Science* 64 (3): 589–608.

Ramsey, Grant, and Charles H. Pence. 2016. "EvoText: A New Tool for Analyzing the Biological Sciences." *Studies in History and Philosophy of Science Part C: Studies in History and Philosophy of Biological and Biomedical Sciences* 57:83–87.

Reeve, Hudson Kern, and Paul W. Sherman. 1993. "Adaptation and the Goals of Evolutionary Research." *Quarterly Review of Biology* 68 (1): 1–32.

Reichenbach, Hans. 1949. *The Theory of Probability.* University of California Press.

Reydon, Thomas A. C. 2021. Misconceptions, Conceptual Pluralism, and Conceptual Toolkits: Bringing the Philosophy of Science to the Teaching of Evolution. *European Journal for Philosophy of Science* 11 (2): 48.

Rice, Sean H. 2004. *Evolutionary Theory: Mathematical and Conceptual Foundations.* Sinauer Associates.

———. 2012. "The Place of Development in Mathematical Evolutionary Theory." *Journal of Experimental Zoology Part B: Molecular and Developmental Evolution* 318 (6): 480–88.

Richerson, Peter J., and Robert Boyd. 2005. *Not by Genes Alone.* Oxford University Press.

Ringsby, Thor Harald, Bernt-Erik Saether, Jarle Tufto, Henrik Jensen, and Erling Johan Solberg. 2002. "Asynchronous Spatiotemporal Demography of a House Sparrow Metapopulation in a Correlated Environment." *Ecology* 83 (2): 561–69.

Roberts, John T. 2016. "The Range Conception of Probability and the Input Problem." *Journal of General Philosophy of Science* 47 (1): 171–88.

Robertson, Alan. 1966. "A Mathematical Model of the Culling Process in Dairy Cattle." *Animal Production* 8:95–108.

Roff, Derek A. 1997. *Evolutionary Quantitative Genetics.* Chapman & Hall.

Rohwer, Yasha, and Collin Rice. 2013. "Hypothetical Pattern Idealization and Explanatory Models." *Philosophy of Science* 80 (3): 334–55.

Romeijn, Jan-Willem. 2017. "Philosophy of Statistics." In *The Stanford Encyclopedia of Philosophy* (Spring 2017), edited by Edward N. Zalta. https://plato.stanford.edu/archives/spr2017/entries/statistics/.

Rosenberg, Alexander, and Frédéric Bouchard. 2005. "Matthen and Ariew's Obituary for Fitness: Reports of Its Death Have Been Greatly Exaggerated." *Biology & Philosophy* 20:343–53.

———. 2015. "Fitness." In *The Stanford Encyclopedia of Philosophy* (Fall 2015), edited by Edward N. Zalta. https://plato.stanford.edu/archives/fall2015/entries/fitness/.

Rosenberg, Noah A. 2006. "Standardized Subsets of the HGDP-CEPH Human Genome Diversity Cell Line Panel, Accounting for Atypical and Duplicated Samples and Pairs of Close Relatives." *Annals of Human Genetics* 70 (6): 841–47.

Rosenthal, Jacob. 2010. "The Natural-Range Conception of Probability." In *Time, Chance, and Reduction: Philosophical Aspects of Statistical Mechanics*, edited by Gerhard Ernst and Andreas Hüttemann, 71–90. Cambridge University Press.

———. 2012. "Probabilities as Ratios of Ranges in Initial-State Spaces." *Journal of Logic, Language, and Inference* 21:217–36.

———. 2016. "Johannes von Kries's Range Conception, the Method of Arbitrary Functions, and Related Modern Approaches to Probability." *Journal of General Philosophy of Science* 47 (1): 151–70.

Rosenthal, Robert. 1979. "The 'File Drawer Problem' and Tolerance for Null Results." *Psychological Bulletin* 86 (3): 638–41.

Roughgarden, J. 1979. *Theory of Population Genetics and Evolutionary Ecology: An Introduction.* Macmillan.

Royall, Richard. 2004. "The Likelihood Paradigm for Statistical Evidence." In *The Nature of Scientific Evidence*, edited by Mark L. Taper and Subhash R. Lele, 119–38. University of Chicago Press.

Ruben, Adam. 2017. *Pinball Wizards: Jackpots, Drains, and the Cult of the Silver Ball.* Chicago Review Press.

Salmon, Wesley C. 1979. "Propensities: A Discussion Review." *Erkenntnis* 14:183–216.

Samuk, Kieran, Jan Xue, and Diana J. Rennision. 2018. "Exposure to Predators Does Not Lead to the Evolution of Larger Brains in Experimental Populations of Threespine Stickleback." *Evolution* 72 (4): 916–29.

Santos, J., M. Pascual, I. Fragata, P. Simões, M. A. Santos, M. Lima, A. Marques, et al. 2016a. "Tracking Changes in Chromosomal Arrangements and Their Genetic Content During Adaptation." *Journal of Evolutionary Biology* 29 (6): 1151–67.

———. 2016b. "Data from: Tracking Changes in Chromosomal Arrangements and Their Genetic Content During Adaptation." Dryad Digital Repository. DOI: 10.5061/dryad.c5bf7.

Santos, J., M. Pascual, P. Simões, I. Fragata, M. Lima, B. Kellen, M. Santos, A. Marques, M. R. Rose, and M. Matos. 2012. "From Nature to the Laboratory: The Impact of Founder Effects on Adaptation." *Journal of Evolutionary Biology* 25 (12): 2607–22.

Schaffer, Jonathan. 2007. "Deterministic Chance?" *British Journal for the Philosophy of Science* 58 (2): 113–40.

Scriven, Michael. 1959. "Explanation and Prediction in Evolutionary Theory." *Science* 130:477–82.

Shaw, Ruth G., Charles J. Geyer, Stuart Wagenius, Helen H. Hangelbroek, and Julie R. Etterson. 2008. "Unifying Life-History Analyses for Inference of Fitness and Population Growth." *The American Naturalist* 172 (1): E35–47.

Simões, Pedro, Marta Pascual, Josiane Santos, Michael R. Rose, and Margarida Matos. 2008. "Evolutionary Dynamics of Molecular Markers during Local Adaptation: A Case Study in *Drosophila subobscura.*" *BMC Evolutionary Biology* 8 (1): 66.

Simões, Pedro, Josiane Santos, Inês Fragata, Laurence D. Mueller, Michael R. Rose, and Margarida Matos. 2008. "How Repeatable Is Adaptive Evolution? The Role of Geographical Origin and Founder Effects in Laboratory Adaptation." *Evolution* 62 (8): 1817–29.

Simon, Herbert A. 1996. *The Sciences of the Artificial*. 3rd ed. MIT Press.

———. 2002. "Near Decomposability and the Speed of Evolution." *Industrial and Corporate Change* 11 (3): 587–99.

Skipper, Robert A., Jr., and Roberta L. Millstein. 2005. "Thinking about Evolutionary Mechanisms: Natural Selection." *Studies in History and Philosophy of Science Part C: Studies in History and Philosophy of Biological and Biomedical Sciences* 36 (2): 327–47.

Slatkin, Montgomery. 1974. "Hedging One's Evolutionary Bets." *Nature* 250 (5469): 704–5.

Sober, Elliott. 1984. *The Nature of Selection*. MIT Press.

———. 1988. "Apportioning Causal Responsibility." *The Journal of Philosophy* 85 (6): 303–18.

———. 2000. *Philosophy of Biology*. 2nd ed. Westview Press.

———. 2001. "The Two Faces of Fitness." In *Thinking about Evolution*, edited by Rama S. Singh, Costas B. Krimbas, Diane B. Paul, and John Beatty, 309–21. Cambridge University Press.

———. 2008. *Evidence and Evolution: The Logic Behind the Science*. Cambridge University Press.

———. 2010. "Evolutionary Theory and the Reality of Macro Probabilities." In *Probability in Science*, edited by Ellery Eells and James Fetzer, 133–61. Springer.

———. 2013. "Trait Fitness Is Not a Propensity, but Fitness Variation Is." *Studies in History and Philosophy of Science Part C: Studies in History and Philosophy of Biological and Biomedical Sciences* 44:336–41.

———. 2020. "Fitness and the Twins." *Philosophy, Theory, and Practice in Biology* 12 (1): 1–13.

Speakman, John R. 2013. "Evolutionary Perspectives on the Obesity Epidemic: Adaptive, Maladaptive, and Neutral Viewpoints." *Annual Review of Nutrition* 33 (1): 289–317.

Spencer, Quayshawn. 2016. "Do Humans Have Continental Populations?" *Philosophy of Science* 83 (5): 791–802.

Sprenger, Jan. 2016a. "Bayesianism vs. Frequentism in Statistical Inference." In *Handbook of the Philosophy of Probability*, edited by Alan Hájek and Chris Hitchcock, 382–405. Oxford University Press.

———. 2016b. "The Probabilistic No Miracles Argument." *European Journal for Philosophy of Science* 6 (2): 173–89.

Sprenger, Jan, and Stephan Hartmann. 2019. *Bayesian Philosophy of Science*. Oxford University Press.

Stearns, Stephen C. 1976. "Life-History Tactics: A Review of the Ideas." *Quarterly Review of Biology* 51 (1): 3–47.

———. 1992. *The Evolution of Life Histories*. Oxford University Press.

Stearns, Stephen C., and Richard E. Crandall. 1981. "Quantitative Predictions of Delayed Maturity." *Evolution* 35 (3): 455–63.

Steele, Richard H. 1986a. "Courtship Feeding in *Drosophila subobscura*. I. The Nutritional Significance of Courtship Feeding." *Animal Behaviour* 34 (4): 1087–98.

———. 1986b. "Courtship Feeding in *Drosophila subobscura*. II. Courtship Feeding by Males Influences Female Mate Choice." *Animal Behaviour* 34 (4): 1099–1108.

Stegenga, Jacob. 2010. "'Population' Is Not a Natural Kind of Kinds." *Biological Theory* 5 (2): 154–60.

———. 2016. "Population Pluralism and Natural Selection." *The British Journal for the Philosophy of Science* 67 (1): 1–29.

Stephens, Christopher. 2004. "Selection, Drift, and the 'Forces' of Evolution." *Philosophy of Science* 71:550–70.

———. 2010. "Forces and Causes in Evolutionary Theory." *Philosophy of Science* 77 (5): 716–27.

Sterelny, Kim, and Philip Kitcher. 1988. "The Return of the Gene." *Journal of Philosophy* 85:339–61.

Sterrett, Susan G. 2014. "The Morals of Model-Making." *Studies in History and Philosophy of Science Part A* 46:31–45.

Strevens, Michael. 1998. "Inferring Probabilities from Symmetries." *Noûs* 32:231–46.

———. 2003. *Bigger than Chaos: Understanding Complexity through Probability.* Harvard University Press.

———. 2005. "How Are the Sciences of Complex Systems Possible?" *Philosophy of Science* 72 (4): 531–56.

———. 2008. *Depth: An Account of Scientific Explanation.* Harvard University Press.

———. 2011. "Probability Out of Determinism." In *Probabilities in Physics*, edited by Claus Beisbart and Stephan Hartmann, 339–64. Oxford University Press.

———. 2013. *Tychomancy: Inferring Probability from Causal Structure.* Harvard University Press.

———. 2015. "Stochastic Independence and Causal Connection." *Erkenntnis* 80 (S3): 605–27.

———. 2016. "The Reference Class Problem in Evolutionary Biology: Distinguishing Selection from Drift." In *Chance in Evolution*, edited by Grant Ramsey and Charles H. Pence, 145–175. University of Chicago Press.

Stroebe, Wolfgang, Tom Postmes, and Russell Spears. 2012. "Scientific Misconduct and the Myth of Self-Correction in Science." *Perspectives on Psychological Science* 7 (6): 670–88.

Suárez, Mauricio, ed. 2009. *Fictions in Science: Philosophical Essays on Modeling and Idealization.* Routledge.

———. 2013. "Propensities and Pragmatism." *Journal of Philosophy* 110 (2): 61–92.

———. 2017a. "The Chances of Propensities." *The British Journal for the Philosophy of Science* 69 (4): 1155–77.

———. 2017b. "Propensities, Probabilities, and Experimental Statistics." In *EPSA15 Selected Papers*, edited by Michela Massimi, Jan-Willem Romeijn, and Gerhard Schurz, 335–45. Vol. 5 of *European Studies in Philosophy of Science*. Springer.

———. 2020. *Philosophy of Probability and Statistical Modeling.* Cambridge University Press.

———. 2022. "The Complex Nexus of Evolutionary Fitness." *European Journal for Philosophy of Science* 12 (1): 9.

Suppes, Patrick, and Mario Zanotti. 1996. *Foundations of Probability with Applications.* Cambridge University Press.

Takacs, Peter, and Pierrick Bourrat. 2021. "Fitness: Static or Dynamic?" *European Journal for Philosophy of Science* 11 (4): 112.

———. 2022. "The Arithmetic Mean of What? A Cautionary Tale about the Use of the Geometric Mean as a Measure of Fitness." *Biology & Philosophy* 37:12.

Tenaillon, Olivier, Jeffrey E. Barrick, Noah Ribeck, Daniel E. Deatherage, Jeffrey L. Blanchard, Aurko Dasgupta, Gabriel C. Wu, et al. 2016. "Tempo and Mode of Genome Evolution in a 50,000-Generation Experiment." *Nature* 536: 165–70.

Thoday, J. M. 1953. "Components of Fitness." *Symposium for the Society for Experimental Biology* 7:96–113.

van Inwagen, Peter, and Meghan Sullivan. 2020. "Metaphysics." In *The Stanford Encyclopedia of Philosophy* (Spring 2020), edited by Edward N. Zalta. https://plato.stanford.edu/archives/spr2020/entries/metaphysics/.

Voight, Benjamin F., Sridhar Kudaravalli, Xiaoquan Wen, and Jonathan K. Pritchard. 2006. "A Map of Recent Positive Selection in the Human Genome." *PLOS Biology* 4 (3): 446–58.

von Kries, Johannes. 1886. *Die Principien Der Wahrscheinlichkeitsrechnung*. Mohr Siebeck.

von Plato, Jan. 1982. "Probability and Determinism." *Philosophy of Science* 49:51–66.

———. 1983. "The Method of Arbitrary Functions." *British Journal for the Philosophy of Science* 34:37–47.

———. 1994. *Creating Modern Probability*. Cambridge University Press.

Wagner, Andreas. 2005. *Robustness and Evolvability in Living Systems*. Princeton University Press.

Wagner, Günter P. 2010. "The Measurement Theory of Fitness." *Evolution* 64 (5): 1358–1376.

Waismann, Friedrich. 1945. "Symposium: Verifiability, Part II." *Proceedings of the Aristotelian Society, Supplementary Volumes* 19:119–50.

Wakeley, John. 2009. *Coalescent Theory: An Introduction*. Roberts and Company.

Walley, Peter, and Terrence L. Fine. 1982. "Towards a Frequentist Theory of Upper and Lower Probability." *The Annals of Statistics* 10 (3): 741–61.

Walsh, Denis M. 2000. "Chasing Shadows: Natural Selection and Adaptation." *Studies in History and Philosophy of Biological and Biomedical Sciences* 31 (1): 135–53.

———. 2007. "The Pomp of Superfluous Causes: The Interpretation of Evolutionary Theory." *Philosophy of Science* 74:281–303.

———. 2010. "Not a Sure Thing: Fitness, Probability, and Causation." *Philosophy of Science* 77 (2): 141–71.

———. 2013. "Descriptions and Models: Some Responses to Abrams." *Studies in History and Philosophy of Science Part C: Studies in History and Philosophy of Biological and Biomedical Sciences* 44 (3): 302–8.

———. 2015a. *Organisms, Agency, and Evolution*. Cambridge University Press.

———. 2015b. "Variance, Invariance and Statistical Explanation." *Erkenntnis*. 80 (S3): 469–89.

Walsh, Denis M., André Ariew, and Mohan Matthen. 2017. "Four Pillars of Statisticalism." *Philosophy, Theory, and Practice in Biology* 9 (1): 1–18.

Waters, C. Kenneth. 2014. "Shifting Attention from Theory to Practice in Philosophy of Biology." In *New Directions in the Philosophy of Science*, edited by Maria Carla Galavotti, Dennis Dieks, Wenceslao J. Gonzalez, Stephan Hartmann, Thomas Uebel, and Marcel Weber, 121–39. Springer.

———. 2017. "No General Structure." In *Metaphysics and the Philosophy of Science: New Essays*, edited by Matthew H. Slater and Zanja Yudell, 81–107. Oxford University Press.

———. 2019. "Presidential Address, PSA 2016: An Epistemology of Scientific Practice." *Philosophy of Science* 86 (4): 585–611.

Waxman, David. 2009. "Fixation at a Locus with Multiple Alleles: Structure and Solution of the Wright Fisher Model." *Journal of Theoretical Biology* 257:245–51.

Weinreich, Daniel M., Richard A. Watson, and Lin Chao. 2005. "Perspective: Sign Epistasis and Genetic Constraint on Evolutionary Trajectories." *Evolution* 59 (6): 1165–74.

Weir, Bruce S. 1996. *Genetic Data Analysis II: Methods for Discrete Population Genetic Data*. Sinauer.

———. 2012. "The Estimation of *F*-Statistics: A Historical View." *Philosophy of Science* 79 (5): 637–43.

Weir, Bruce S., and C. Clark Cockerham. 1984. "Estimating *F*-Statistics for the Analysis of Population Structure." *Evolution* 38 (6): 1358–70.

Weir, Bruce S., and Jérôme Goudet. 2017. "A Unified Characterization of Population Structure and Relatedness." *Genetics* 206 (4): 2085–2103.

Weir, Bruce S., and W. G. Hill. 2002. "Estimating *F*-Statistics." *Annual Review of Genetics* 36:721–50.

Weisberg, Michael. 2013. *Simulation and Similarity: Using Models to Understand the World*. Oxford University Press.

Wells, Jeffrey V., and Milo E. Richmond. 1995. "Populations, Metapopulations, and Species Populations: What Are They and Who Should Care?" *Wildlife Society Bulletin (1973–2006)* 23 (3): 458–62.

Williams, David. 1991. *Probability with Martingales*. Cambridge University Press.

Williams, George C. 1966. *Adaptation and Natural Selection*. Princeton University Press.

Wilson, Mark. 2006. *Wandering Significance: An Essay on Conceptual Behavior*. Oxford University Press.

Wimsatt, William C. 1972. "Teleology and the Logical Structure of Function Statements." *Studies in History and Philosophy of Science* 3 (1): 1–80.

———. 1974. "Complexity and Organization." In *PSA 1972*, edited by Kenneth F. Schaffner and Robert S. Cohen, 67–86. Reidel. (Reprinted in Wimsatt 2007.)

———. 1980a. "Randomness and Perceived Randomness in Evolutionary Biology." *Synthese* 43:287–329.

———. 1980b. "Reductionistic Research Strategies and Their Biases in the Units of Selection Controversy." In *Scientific Discovery*, edited by Thomas Nickles, 213–59. Vol. 2, *Case Studies*. Reidel. (Reprinted in *Conceptual Issues in Evolutionary Biology*, ed. Elliott Sober. MIT Press, 1984.)

———. 1981. "The Units of Selection and the Structure of the Multi-level Genome." In *PSA: Proceedings of the Biennial Meeting of the Philosophy of Science Association* 1980 (2): 122–83.

———. 2002. "Functional Organization, Analogy, and Inference." In *Functions: New Essays in the Philosophy of Psychology and Biology*, edited by André Ariew, Robert Cummins, and Mark Perlman, 173–221. Oxford University Press.

———. 2007. *Re-engineering Philosophy for Limited Beings: Piecewise Approximations to Reality*. Harvard University Press.

Wimsatt, William C., and James R. Griesemer. 2007. "Reproducing Entrenchments to Scaffold Culture: The Central Role of Development in Cultural Evolution." In *Integrating Evolution and Development*, edited by Roger Sansom and Robert N. Brandon, 227–323. MIT Press.

Wiser, Michael J., and Richard E. Lenski. 2015. "A Comparison of Methods to Measure Fitness in *Escherichia coli*." *PLOS ONE* 10 (5): 1–11.

Wiser, Michael J., Noah Ribeck, and Richard E. Lenski. 2013. "Long-Term Dynamics of Adaptation in Asexual Populations." *Science* 342 (6164): 1364–67.

Woodward, James. 2003. *Making Things Happen*. Oxford University Press.

Wright, Sewall. 1934. "The Method of Path Coefficients." *The Annals of Mathematical Statistics* 5 (3): 161–215.

———. 1951. "The Genetical Structure of Populations." *Annals of Eugenics* 15:323–54.

———. 1969. *Evolution and the Genetics of Populations*. Vol. 2 of *The Theory of Gene Frequencies*. University of Chicago Press.

Zynda, Lyle. 2000. "Representation Theorems and Realism about Degrees of Belief." *Philosophy of Science* 67 (1): 45–69.

Index

Page numbers in italics refer to figures and tables.